STRENGTH OF BEAMS

UNDER

TRANSVERSE LOADS.

BY

PROF. W. ALLAN.

FORMERLY OF WASHINGTON AND LEE UNIVERSITY,
LEXINGTON, VA.

NEW YORK:
D. VAN NOSTRAND, PUBLISHER,
23 MURRAY AND 27 WARREN STREET.
1875.

PREFACE.

The usual discussion of stress in beams loaded transversely, involving as it does the calculus, is unintelligible to that large class of builders, and others, whose mathematical training has not extended beyond the elements of conic-sections. Yet, the most important cases may be explained without using the higher mathematics.

The aim of the following pages is to put into brief and convenient shape, without resorting to the higher mathematics, the discussion of the most important and common cases of horizontal beams under vertical loads. In beams of rectangular cross-section the formulæ given are exact under the assumptions generally made in regard to the laws of elasticity. In flanged beams the approximations made use of are those adopted in ordinary practice.

A graphic method of determining the *amount of resistance* is added, which is applicable to all cross sections, and which, if used with ordinary skill and care, will give results exact enough for every purpose. This method is so simple, and of such general application, that it commends itself to practical builders. It is easily understood, and requires no other appliances but those in the hands of every draughtsman. In the case of unsymmetrical cross-sections, (as for instance, iron rails), it affords an easy means of obtaining sufficiently accurate results, without resorting to the tedious calculations otherwise necessary.

This discussion (in which there is no attempt at originality) was prepared for the use of the intermediate classes in Engineering, at Washington and Lee University. Having been found serviceable to them, both during their college career and in subsequent practice, I am induced to hope that it may save time and labor to those engaged in the same pursuits.

In preparing the manuscript for the press, and correcting the proofs, I have been much indebted to one of my former pupils, Mr. Julius Kruttschnitt, C.E., now Instructor in the McDonogh School.

August, 1875.

ERRATA.

Page 14, in fig. 5, N and N,' should change places.

" 19 and 53, figs. 8 and 35 should be turned, so as to make A C vertical.

" 41, third line from top, add (Fig. 26) after C P.'

" 60, near bottom, Wx (l–x) should read $\frac{Wx}{l}$ (l–x)

" 78, second line from top "Fig. (2)" should read (Fig. 52).

" 78, near bottom, read d = 5c. x, instead of d = 5cx.

" 86, fifth line from top, "proportional" should be *proportioned.*

" 102, fourth line from bottom, omit "these."

" 111, second line from bottom, "those" should be *that.*

" 112, second line from bottom, "not" should be *cut*; and in bottom line, "beams" should be *bases.*

STRENGTH OF BEAMS UNDER TRANSVERSE LOADS.

In the following discussion the ordinary cases of loaded beams are treated without resorting to the higher mathematics. It is an attempt to compile from various sources the simplest methods of treating such cases as arise most frequently in practice.

Transverse stress is produced by a load applied to a beam in a direction perpendicular, or inclined, to its length. A D (Fig. 1) is a beam subjected to such a stress.

In this kind of stress a compression of the particles or fibres on one side of the

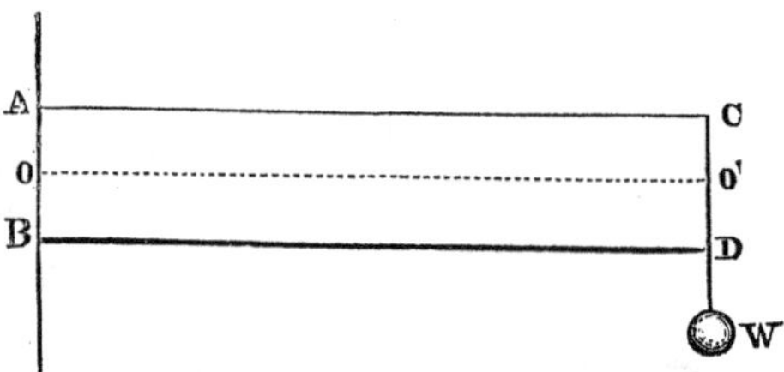

FIG. 1.

beam, and an extension of those on the other, are produced. In consequence of this the beam *bends*. Experiment shows that the amount of compression on the the one side, and of extension on the other, diminishes as we go inwards from the top or bottom towards the centre, and at some intermediate plane, O O′, becomes zero. The fibres at this plane being neither lengthened nor shortened, it is called the *neutral* plane, and its intersection by the plane of vertical section is called the *neutral line or axis*. Experiment also shows that, from this plane towards the top and bottom, the amount of extension and compression may, for the stresses that occur in ordinary practice, be considered as varying directly with the distance from the neutral plane.

The extreme top and bottom fibres suffer the greatest compression and extension, and in case of rupture, the rupture begins with them. Some question exists as to the exact location of the neutral line or plane, but for slight deflections it passes through the centre of gravity of the cross section of the beam, and it is very probable that it never deviates from this position.

In discussing transverse stress, the assumptions based upon experiment may be stated as follows :

1. The forces on the fibres are directly as the amount of extension or compression they produce ; (*Ut tensio sic vis,*) and since the extension and compression increase as the distance from the neutral axis, the forces vary in the same proportion.

2. Within *elastic limits* the extension and compression at equal distances from

the neutral axis are equal, and the forces producing them are equal.

3. The *neutral axis* passes through the centre of gravity of the cross section.

RECTANGULAR BEAMS.

Let us now discuss the relations existing between the forces in, and on, transversely loaded rectangular beams, the load being supposed to be vertical in direction and the beam horizontal.

Case I.

Let A D (Fig. 2) be a beam so thin that it may be considered as composed of but one layer of fibres or particles. Let it be fastened in a wall at A B, and be loaded with W′ at the other end. Neglect for the time the weight of the thin beam itself, which is small. Imagine it to be cut by a vertical plane E F at any point, and let us see under what forces the part E D is held in equilibrium.

The only external force on E D is W′ acting at D downwards, and E D is pre-

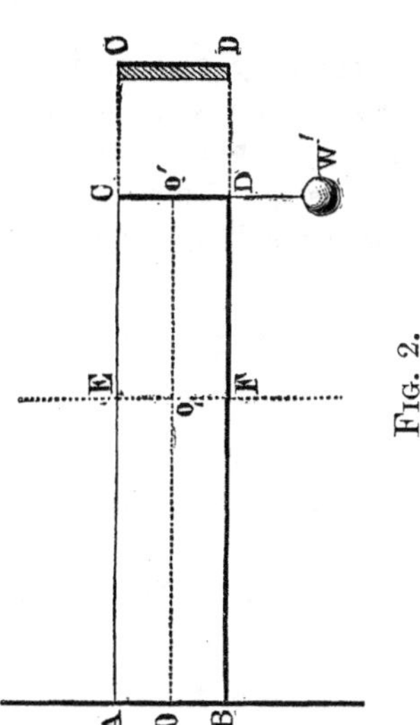

Fig. 2.

vented from falling under this weight by the *resistance of the fibres* at E F. To analyze these forces, let us take O_1 as an origin of coordinates, and $O_1 O'$ as the axis of x, and $O_1 E$ as the axis of y, and as the forces are all in one plane, find their components along these axes. The *internal* forces, or resistance of the fibres at E F, are:

1. The *horizontal forces* which are tensile above and compressive below, and which increase from zero at O_1 just in proportion as we go from that point towards the upper or lower edge of the beam. (Fig. 3.)

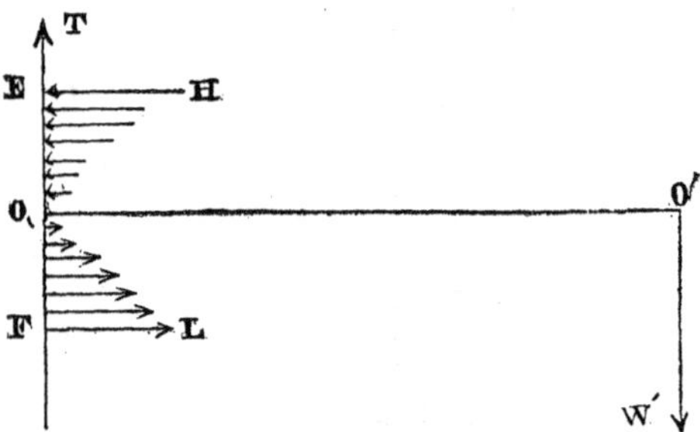

Fig. 3.

2. The *vertical force.* This is called the *shearing force*, or *transverse* shearing force. It resists the tendency of the part of the beam E D to slide down on the surface E F. The existence of this force may be realized if we conceive the beam to be divided into two parts by the vertical plane E F, and those parts to be united by some very elastic substance,

as india-rubber. Then the beam would take the form shown in Fig. 4, the part

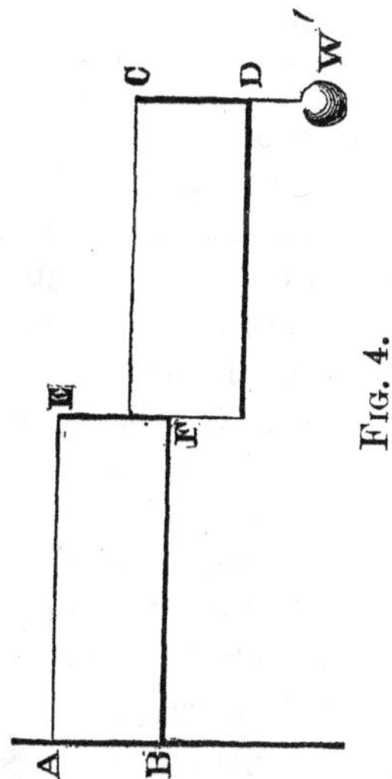

Fig. 4.

F C sliding down on the other. The force in the beam that resists this sliding is represented in Fig. 3, by the vertical arrow at E. Let it be called T. In Fig. 3 are represented all the forces we have to deal with. Since this system of forces is balanced, the following equations must be fulfilled:

$$\Sigma X=0 \quad \Sigma Y=0 \quad \Sigma M=0 \qquad (1)$$

That is: the sum of all the horizontal forces (ΣX), and the sum of all the vertical forces (ΣY) must each be equal to zero, and the sum of the moments about any point as O_1 must also equal zero.

The only horizontal forces in the system are the two triangular groups of forces $E O_1 H$ and $F O_1 L$, representing the sum of the tensile and compressive stresses on the fibres. As the group $E O_1 H$ acts in a direction opposite to that of the group $F O_1 L$, and as the algebraic sum of the two groups is zero ($\Sigma X=0$), the groups must be equal to each other. This is indicated in the figure by the equality of the triangles $E O_1 H$ and $F O_1 L$.

The vertical forces are W' and the shearing force T at E F, and since

$$\Sigma Y=T-W'=0. \quad \text{We have}$$

$$T=W' \qquad (2)$$

Next obtain the moments of all the forces about O_1 and place the sum of these moments = zero. Replace the tensile and compressive forces by their resultants. The resultant or sum of all the tensile forces represented by the triangle $E O_1 H$ (Fig. 3) may evidently be represented by the area of the triangle of which the base $E O_1$ is the distance over which the forces are distributed, and the altitude E H is the stress in the outside fibre. Let

$$S = \text{this stress} = EH$$

and $d = EF = \text{depth of beam}$

Then area $E O_1 H = \frac{1}{4} S d = N$ = resultant of tensile forces. Similarly

Area $F O_1 L = \frac{1}{4} S d = N'$ = resultant of compressive forces.

These resultants will pass through the centres of gravity of the triangles $E O_1 H$ and $O_1 F L$, since the little forces of which they are composed are represented by these triangles. Hence the direction of N will intersect $O_1 E$ at a point G (Fig. 5), whose distance from O_1

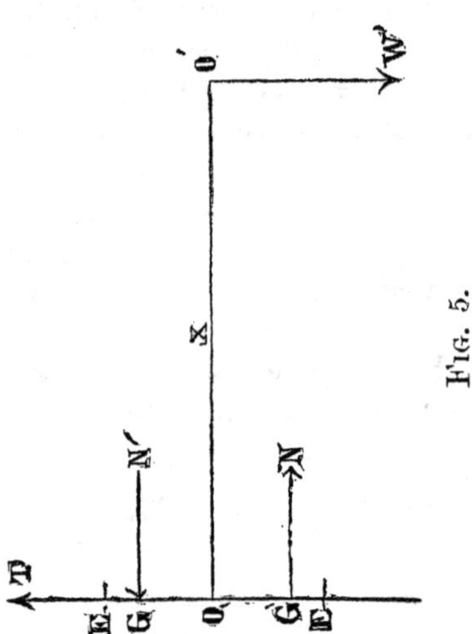

Fig. 5.

is $= \frac{2}{3}\,E\,O_1 = \frac{2}{3}\cdot\frac{d}{2} = \frac{d}{3}$. This is the lever arm of *N* about O_1. That of *N'* is $G'O_1$ also $= \frac{d}{3}$. Hence the sum of the moments of these two forces about O_1 (since they both tend to produce left-handed rotation) is

$$-N\,\frac{d}{3} - N'\,\frac{d}{3} = -\frac{1}{6}\,S\,d^2$$

The force T since its direction passes through O_1 has no lever-arm, and hence its moment is zero.

If the distance from O' to O_1 be called x, the moment of the weight W' is $= + W'x$ (since it tends to produce right-handed rotation)

$$\therefore W'x - \frac{1}{6} S d^2 = \Sigma M = 0$$

$$\therefore W'x = \frac{1}{6} S d^2 \qquad (3)$$

So far we have considered a beam whose breadth is that of only one row of fibres, but a beam of any breadth may be made up of a number of such slices placed side by side, and if $b =$ the number of slices, or breadth of the beam, and $W =$ the weight hung at the end of it, then eq. (3) becomes

$$W x = \frac{1}{6} S b d^2 \qquad (4)$$

This discussion is general and will apply to any section as well as to E F. S and x are the variables in eq. (4), and these quantities will have different values at the different sections, which values increase as we go towards A B (Fig. 2), but the form of the equation will evidently be unchanged. If A C (Fig. 2) be $= l$, we have for the section at A B (Fig. 2)

$$W\,l = \frac{1}{6}\,S\,b\,d^2 \qquad (5)$$

A B is the section of greatest stress, and the beam if overloaded will break there.

The quantity $\frac{1}{6}\,S\,b\,d^2$ is called the *moment of resistance* of the fibres, or moment of the internal forces, and is often written *M* for brevity. *Wx* is called the moment of the weight, or moment of the external forces. Let the maximum value of $\frac{1}{6}\,S\,b\,d^2$ (eq. 5) be called M_0.

We may illustrate geometrically the variation of the moments $M = W\,x$, and consequently of the stresses produced on the outside fibres from A to C.

In Fig. 6 let A C be the beam. Take a line on some scale to represent the

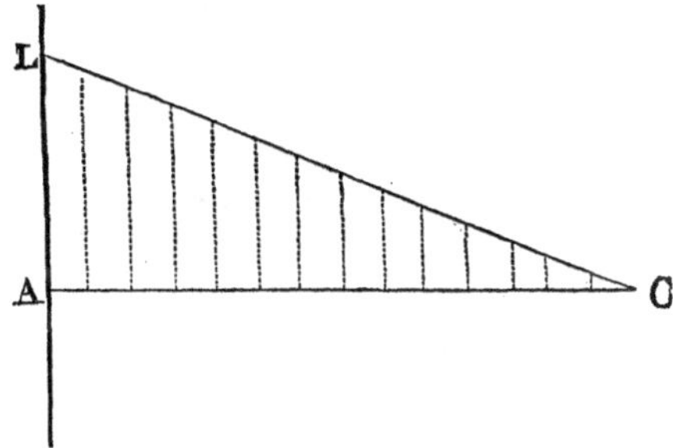

FIG. 6.

value of $M_0 = W l$, and lay it off from A perpendicular to A C. Let A L be this line. Draw L C. Then the dotted perpendiculars in the triangle L A C will represent the moments of resistance in the beam at the several points at which they are drawn.

From eq. (2) it is seen that the shearing force is constant at every section of the beam. This force we may assume with sufficient accuracy, for our present purpose, to be uniformly distributed over the cross section of the beam on which it acts. Hence if A = area of cross section, and $t=$ shearing force on a unit of the surface,

$$T=W=At \qquad (6)$$

Lay off AC (Fig. 7) $=l$ and CP=W.

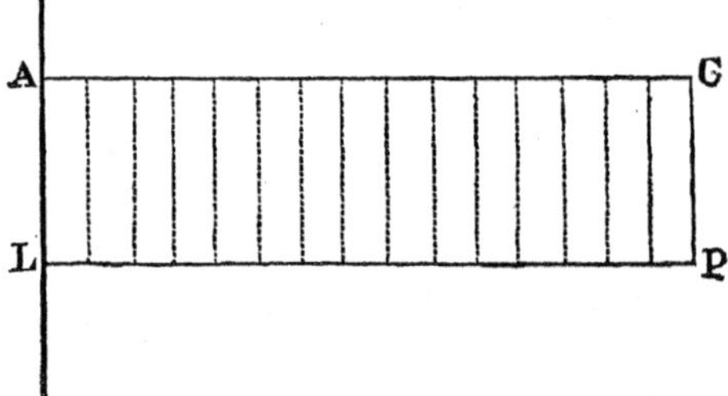

FIG. 7.

Then the rectangle AP represents geometrically the shearing stress at every point of the beam.

Corollary. When several weights as W W_1 W_2 (Fig. 8) are suspended from the beam at different points, the moment of resistance at any point is equal to the joint moments of the weights at that point. Thus, calling distances measured from C, G, and E towards A, x, x_1 and x_2

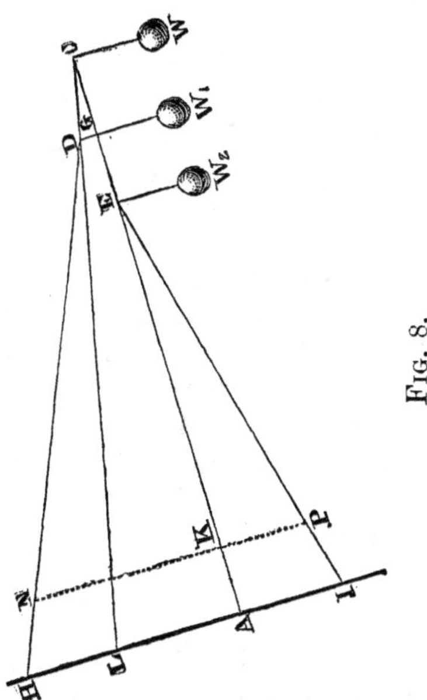

Fig. 8.

respectively, we have for the equation of moments for points between C and G

$$W\,x=\frac{1}{6}\,S\,bd^2$$ Between G and E

$$W\,x+W_1\,x_1=\frac{1}{6}\,S\,b\,d^2$$ While at K,

for instance, it is

$$W x + W_1 x_1 + W_2 x_2 = \frac{1}{6} S b d^2 \qquad (7)$$

The shearing force at K is

$$T = W + W_1 + W_2 \qquad (8)$$

Geometrically. Let A C (Fig. 8) = l A G=l_1 and A E=l_2. Lay off A L=W l, L H=$W_1 l_1$ and A I=$W_2 l_2$. Draw the triangles as in Fig. 8. Then N P=total moment at K, for instance.

The shearing force is represented by the rectangles A P, N O, and S T (Fig. 9), and at any point in the beam is equal to the sum of the weights between that point and C.

EXAMPLES.

(1.) Suppose the safe stress per square inch to be 1,000 lbs. (= S), and l = 10 ft., b=3 inches, and d=12 inches, what weight will the beam support ?

(2.) Suppose W=$\frac{3}{4}$ ton, l=12 ft., b=2 inches, what must be the depth (d) of a rectangular cast iron beam, so that S shall not exceed 4 tons ?

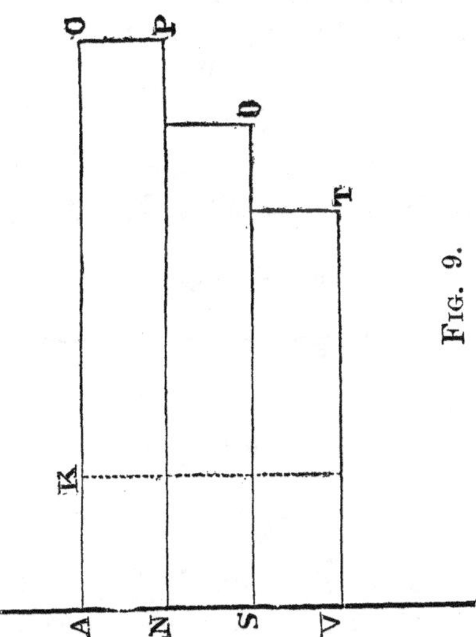

Fig. 9.

Case II.

Let the beam be as in the last case, but with the load distributed uniformly over it (Fig 10). Let w=weight on a unit of length, W=total weight on A C, l=A C=length, d=A B=depth, x=E C as before. Then the forces to be considered are represented in Fig 11, the little arrows along E C representing the weight distributed along the beam.

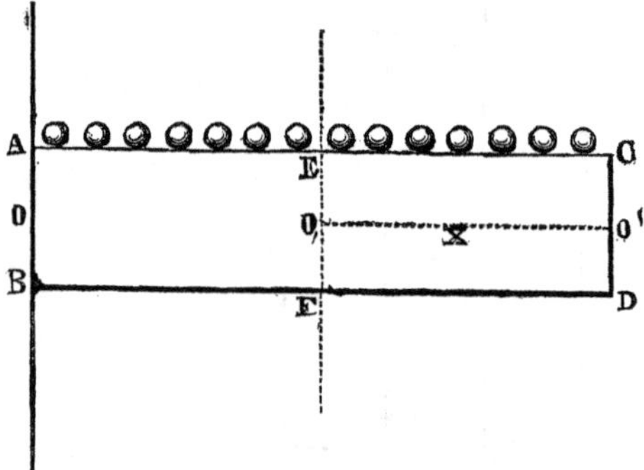

FIG. 10.

Replace the weights along E C by their resultant, which is $= wx$, and which should be applied at the middle point of E C, since the little weights on the beam are uniformly distributed. Then putting the resultants N and N′ in place of the tensile and compressive force, and proceeding as before, we have

$$T - w\,x = 0$$

$$T = w\,x \qquad (9)$$

Also $$w\,x.\frac{x}{2}-\frac{1}{6}\,S\,b\,d^2 \;=0$$

$$\frac{w\,x^2}{2}=\frac{1}{6}\,S\,b\,d^2 \;=M \quad (10)$$

At A B (Fig. 10) these equations become

$$\left.\begin{array}{l} T_0=w\,l=W \\ \frac{w\,l^2}{2}=\frac{W\,l}{2}=\frac{1}{6}\,S\,b\,d^2=M_0 \end{array}\right\} \quad (11)$$

By comparing the last equation with eq. (5), we see that if the weight and beam be the same the stress on the fibres in this last case is only *one-half* what it was in the former, or, what amounts to the same, the beam will bear twice as much distributed over it, as it will when the weight is concentrated at the extremity.

From eq. (9) we see that the shearing force is not constant as in the last case, but varies as x. It is greatest at A.

Geometrically. The equation $M=\frac{1}{2}\,w\,x^2$ corresponds with that of a parabola with vertex at C and axis vertical. Lay off A L (Fig. 12) $=\frac{1}{2}\,w\,l^2$, and through L and C draw a parabola. The ordinates

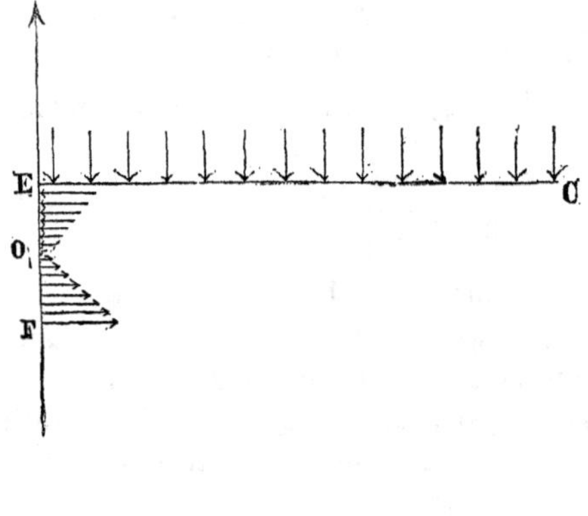

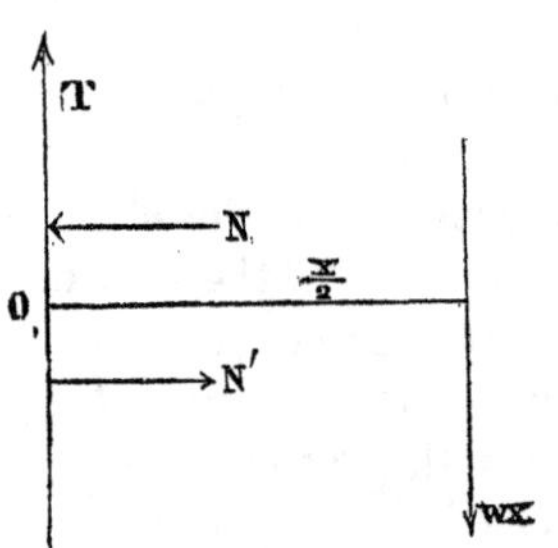

FIG. 11.

of this parabola (dotted in the figure) will represent the moments at the several points.

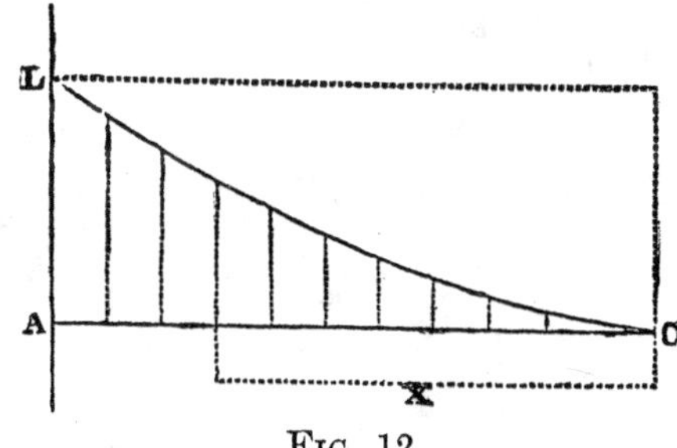

FIG. 12.

The equation $T = w x$ is represented by the triangle A P C (Fig. 13), which therefore gives the shearing stress at every point of the beam when A P is taken $= w\, l$.

Corollary 1. When the load is distributed over only a part of the beam as in Fig. (14), let R C$=m=$the loaded part, and take the other letters as before. Then the equations for any section in the loaded part are evidently the same as those just obtained, viz.:

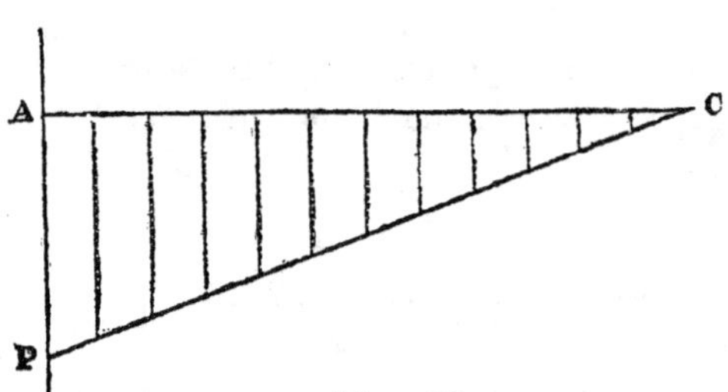

FIG. 13.

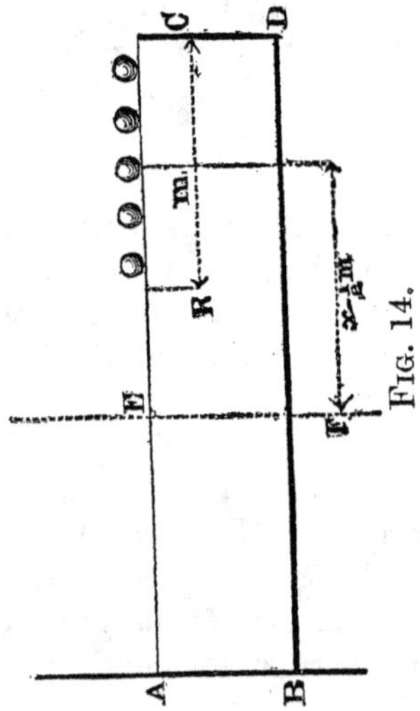

FIG. 14.

$$\left.\begin{array}{ll} & \frac{1}{2}\,w\,x^2 = \frac{1}{6}\,S\,b\,d^2 = M \\ \text{At R} & \frac{1}{2}\,w\,m^2 = M \end{array}\right\} \quad (12)$$

And $\quad T = w\,x$

But at any section E F between A and R the moment of the load is $= w\,m\,(x - \frac{1}{2}\,m)$, the latter factor being the distance from the centre of gravity of the load to the section E F. The moment of resistance having the same form as before, we have for the equation of moments for any section in R A

$$w\,m\,(x - \tfrac{1}{2}\,m) = \frac{1}{6}\,S\,b\,d^2 = M \quad (13)$$

At A this becomes

$w\,m\,(l - \frac{1}{2}\,m) = M_0$ the greatest moment.

The shearing force at E F being equal in amount and opposite in direction to the whole load between E F and C will be

$$T = w\,m \quad (14)$$

Geometrically. For the moments: lay off $A\,C = l$ and $C\,R = m$ (Fig. 15). At R erect $D\,R = \frac{w\,m^2}{2}$, and at A make A L

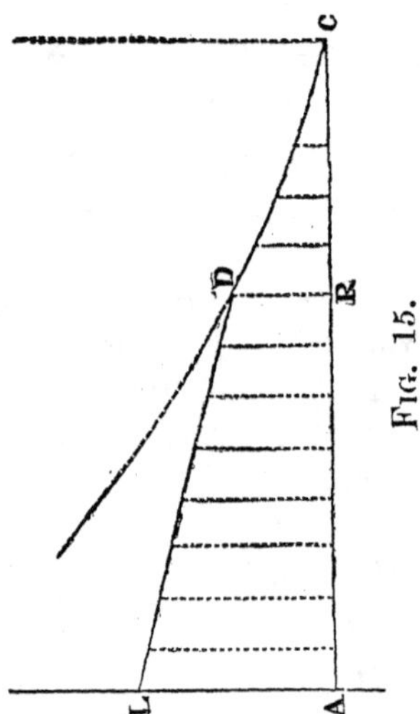

Fig. 15.

$=M_0$. Through C and D draw a parabola as in the last case, and (since eq. [13] is of the first degree) through D and L draw a straight line. Then from C to R the ordinates of the parabola represent the moments, and from R to A they are represented by the ordinates of the trapezoid R L. For the shearing stress (Fig. 16), lay off from R, RN=wm.

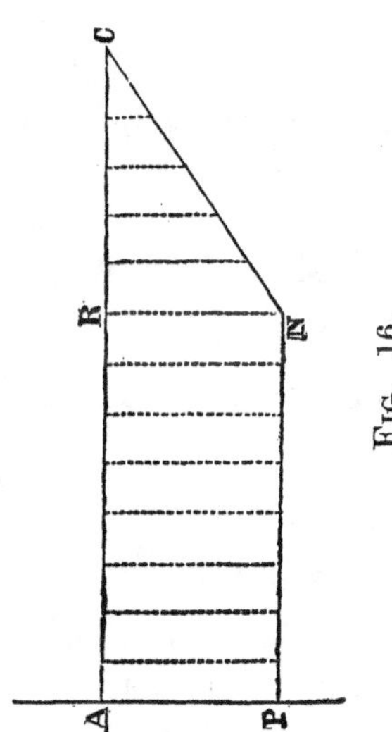

Fig. 16.

Draw C N and N P. Then the triangle C R N (corresponding to the equation $T=w\,x$) gives the shearing force at each point in C R while the rectangle R P (corresponding to the equation $T=w\,m$) gives the force in the remaining segment of the beam.

Corollary 2. When there is a load W at the extremity C, in addition to the load uniformly distributed over the beam (Fig. 17), we have a combination of

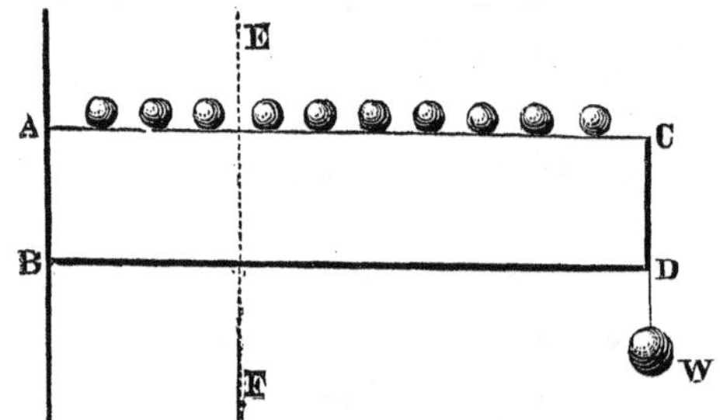

FIG. 17.

Cases I. and II. and the moment of the external forces at any section, E F, is

$$\tfrac{1}{2} w x^2 + W x$$

Hence the equation of moments is

$$M = \frac{1}{6} S b d^2 = \tfrac{1}{2} w x^2 + W x \qquad (15)$$

This is greatest at A, or

$$M_0 = \frac{1}{6} S b d^2 = \tfrac{1}{2} w l^2 + W l \qquad (16)$$

The shearing force

$$T = w x + W \qquad (17)$$

At A

$$T_0 = w l + W \qquad (18)$$

Geometrically. The simplest way of representing the moments is to construct those due to each kind of weight, and then combine them. Thus, let $M' = \frac{wx^2}{2}$ and $M'' = Wx$. Construct M′ as in *Case II.*, it being represented by a parabola with vertex at C and axis vertical, and M″ as in *Case I.*, it being represented by a triangle (placed under A C for convenience). Since $M = M' + M''$ from eq. (15) we have the total moment at any point E (Fig. 18) represented by the sum of the ordinates of these two figures = N P Fig. (18.)

The moments may also be represented by the parabola corresponding to eq. (15) as in Fig. (19.) This parabola has its vertex at C′ and not at C. Of course, only that part of the curve between C and A is applicable to our purpose.

The shearing force is represented by adding the triangle N P P′ Fig. (20) representing the variable part wx of T to the rectangle C P which represents the constant part W of T.

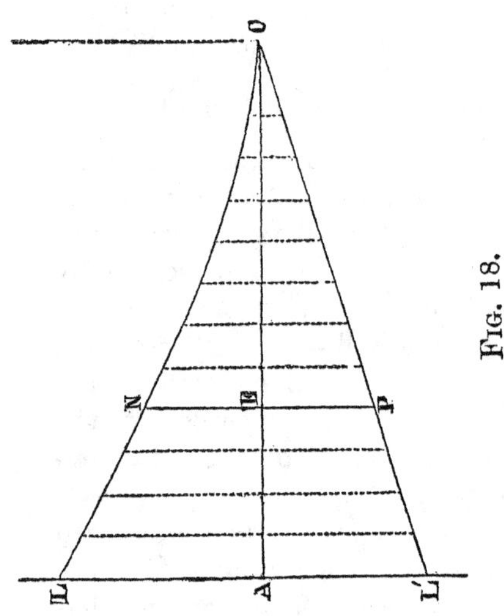

FIG. 18.

EXAMPLE.

Discuss the forces when the load is distributed as shown in Fig. (21) assuming various values for L N and N C as well as for w and W.

Case III.

Let the beam whose length is l rest

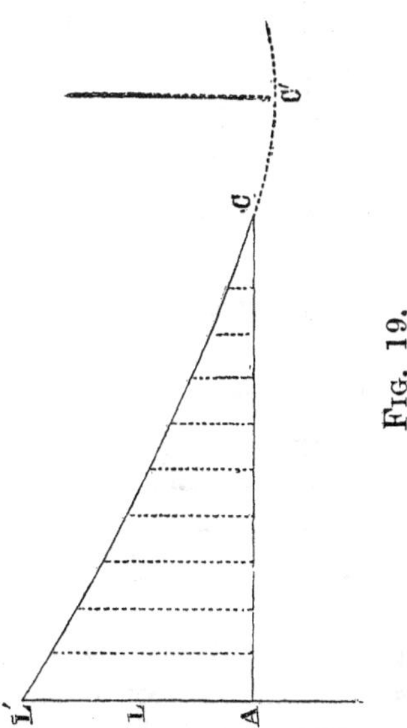

Fig. 19.

upon supports at B and D (Fig. 22) and let it be loaded at some point G with a single weight W. Let m and n be the segments into which the beam is divided at the point G of the application of the weight.

First find the proportions of the

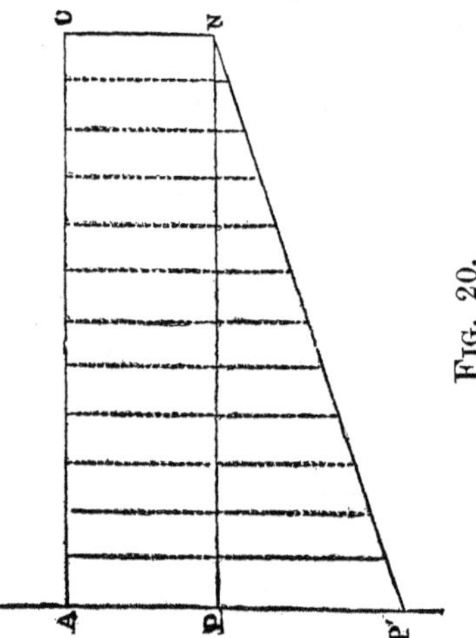

Fig. 20.

weight supported at B and D, or in other words, the reactions of the supports. By the principle of the lever the respective portions of the weight supported at B and D are inversely proportional to the distances of these points from G. Thus, let W′=reaction at D and W″=reaction at B and then

$$W' : W'' :: m : n$$
$$W''+W' : W' :: m+n : m$$

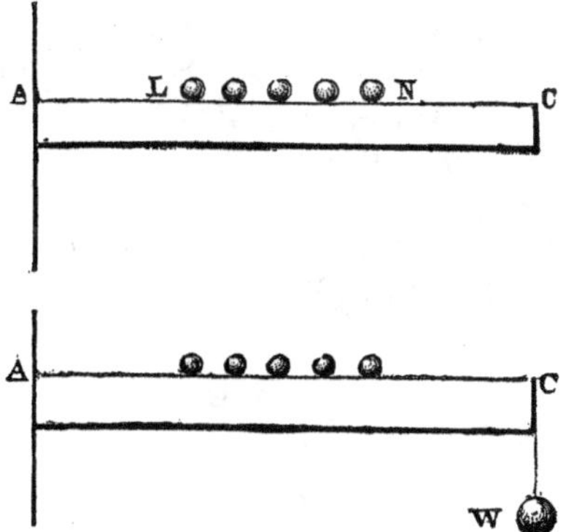

FIG. 21.

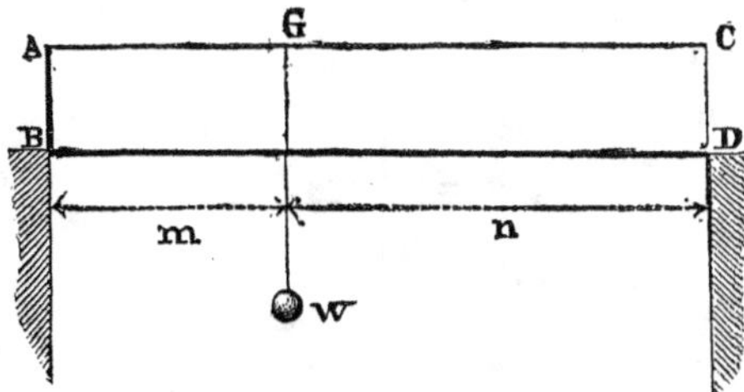

FIG. 22.

But $W' + W'' = W$ and $m + n = l$

$$\left.\begin{aligned} \therefore\ W' &= \frac{m}{l} W \\ \text{And so}\qquad W'' &= \frac{n}{l} W \end{aligned}\right\} \qquad (19)$$

Now apply the conditions of equilibrium to any part A E of the beam counting from A. Fig. (23).

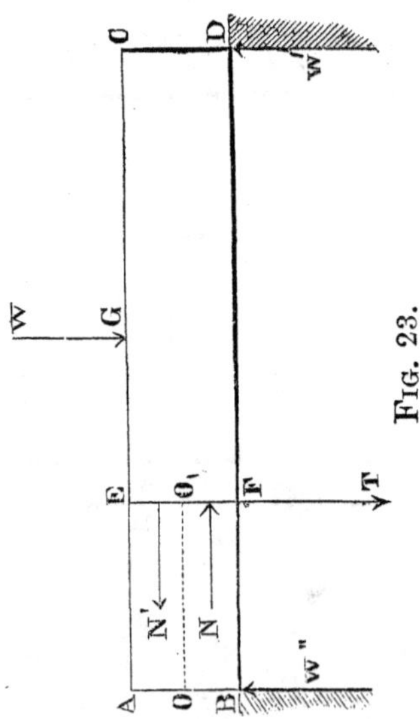

Fig. 23.

1st, Between A and G. $\Sigma X=0$ merely indicates the equality of N and N′, as these are the only horizontal forces. $\Sigma Y=0$ shows that the shearing force at the section E F is downwards

$$\text{and}=W'' \quad \therefore \quad T=W''=\frac{n}{l}W \qquad (20)$$

$\Sigma M=0$. The joint moment of N and N′ is as before$=\frac{1}{6}\,S\,b\,d^2$. That of T is zero. The only other force acting on A E is W″, the reaction of the abutment. Let $O\,O_1=x$. Then the moment of W″ is

$$W''\,x=\frac{n}{l}.\,W\,x$$

$$\text{Hence} \quad \frac{n}{l}.\,W\,x-\frac{1}{6}\,S\,b\,d^2=0$$

$$\therefore \frac{n}{l}\,W\,x=\frac{1}{6}\,S\,b\,d^2=M \quad (21)$$

2d, Between G and C (Fig. 24). Here, between A and E are the two external forces W″ at B and W at G. Hence the shearing force at E F is upwards and

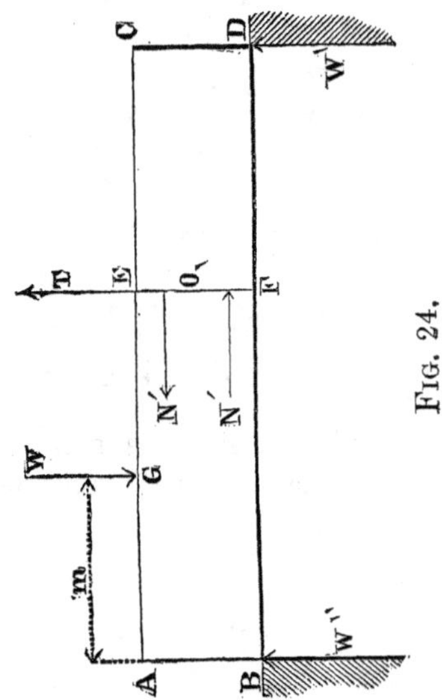

FIG. 24.

$$T = W'' - W = -W' = -\frac{m}{l} W. \qquad (22)$$

For $\Sigma M = 0$ we have

$$-\frac{1}{6} S\, b\, d^2 + W'' x - W\,(x - m) = 0$$

$$\therefore \frac{n}{l} W x - W(x - m) = \frac{1}{6} S\, b d^2 = M \qquad (23)$$

The greatest value of eq. (21) is at G, where it becomes identical with eq. (23) for the same point. This value is

$$\frac{m.\,n}{l}\,W=\frac{1}{6}\,S\,b\,d^2=M_0 \qquad (24)$$

At A and C the moments are zero.

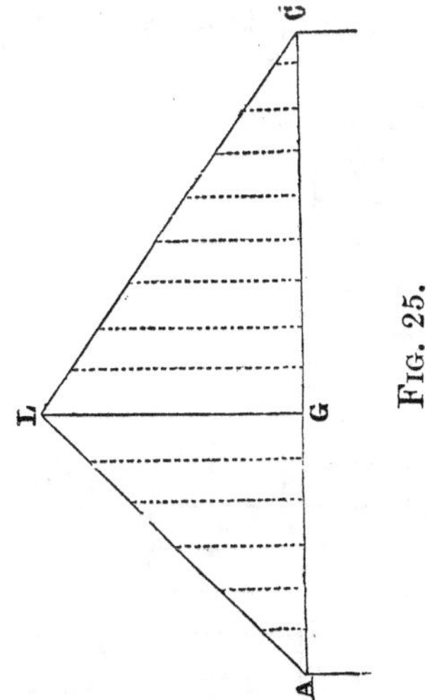

Fig. 25.

Geometrically. From eq. (21), which is of the first degree, it is seen that the moments vary in A G as they did in *Case I.* Hence they may be represented by the ordinates of the triangle A L G. (Fig. 25).

Eq. (23) is also that of a straight line cutting the axis of X at C. Hence the moments in G C are represented by the triangle G L C (Fig. 25). The shearing

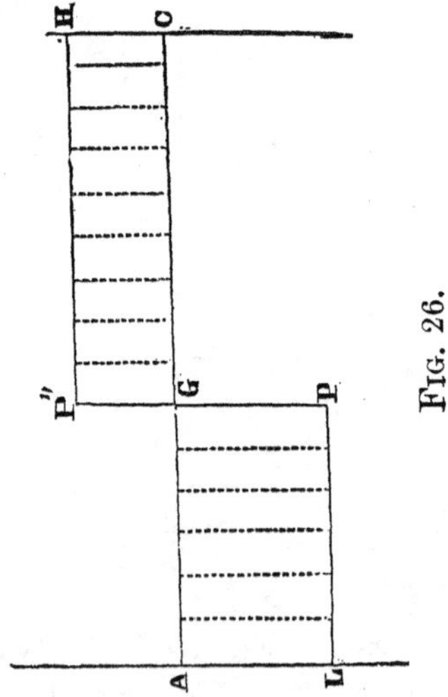

FIG. 26.

force in A G is represented by the rectangle A P and that in G C by the rectangle C P′.

Note, That the maximum moment (at G) corresponds to the point where the shearing force passes through zero.

Corollary 1. When the weight W is at the middle of the beam we have

$$n = m = \tfrac{1}{2} l$$
$$W' = W'' = \tfrac{1}{2} W$$

Then eq. (21) becomes

$$\tfrac{1}{2} W x = \frac{1}{6} S b d^2 = M$$

and (23) is

$$\tfrac{1}{2} W (l - x) = \frac{1}{6} S b d^2 = M \qquad (25)$$

At the centre $M_0 = \tfrac{1}{2} W . \tfrac{1}{2} l = \tfrac{1}{4} W l$ (26)

The triangles A L G and G L C (Fig, 27) represent the moments in this case. G L having been laid off$=\tfrac{1}{4} W l$.

The shearing force throughout the beam is then $T = \tfrac{1}{2} W$ (27) as is shown in the rectangles (Fig. 28).

Comparing the value of M_0 given in

2

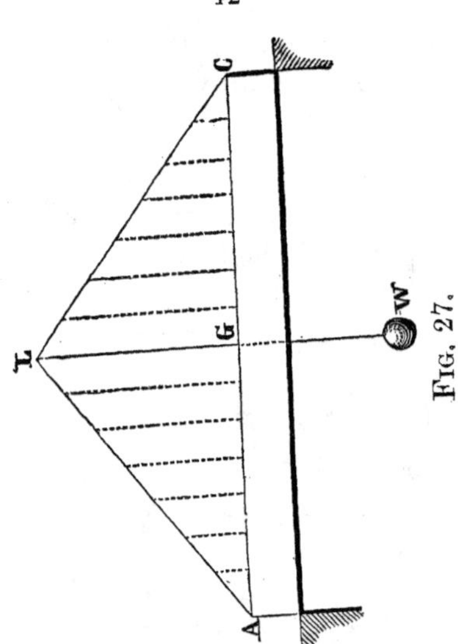

Fig. 27.

eq. (5) *Case I.* with that of M_0 in eq. (26) we see that the load and the length being the same, a beam will bear *four* times as much with both ends supported, and the load placed in the middle as it will do with one end fixed and the other loaded.

Corollary 2. When there are several weights, as in Fig (29), the moment of

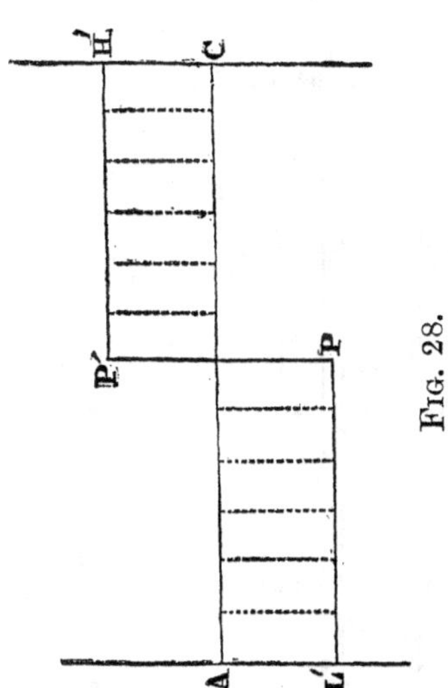

Fig. 28.

the external forces at any section is that due to the action of all the weights. Let the segments into which the weights W, W_1 and W_2 divide the beam be m and n for W, m_1 and n_1 for W_1 and m_2 and n_2 for W_2.

Let R_1=reaction of abutment at B and
R_2=reaction of abutment at D and
l=length and let x be counted from A as before.

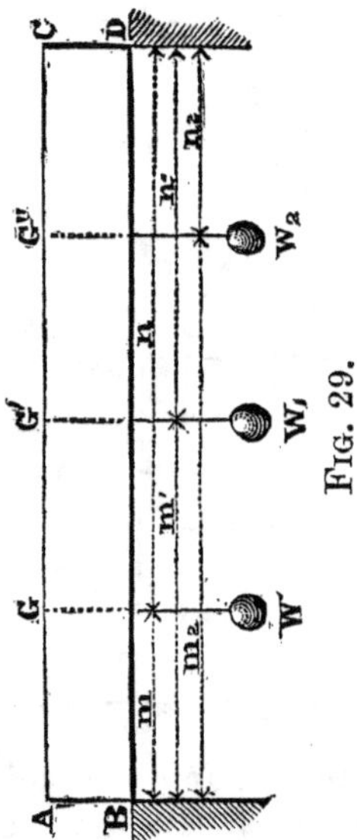

FIG. 29.

The reaction of the abutment at B due to these weights is

$$\left.\begin{aligned} R_1 &= \frac{n}{l}W + \frac{n_1}{l}W_1 + \frac{n_2}{l}W_2 \\ \text{So at D}\quad R_2 &= \frac{m}{l}W + \frac{m_1}{l}W_1 + \frac{m_2}{l}W_2 \end{aligned}\right\} \quad (28)$$

For any section between A and G, R_1

is the only external force and hence the equation of moments is

$$R_1 x = \frac{1}{6} S b d^2 = M \qquad (29)$$

The shearing force is $T = R_1$ (30)

For every section between G and G′ the weight W is to be taken into consideration and the equations are:

$$R_1 x - W(x-m) = \frac{1}{6} S b d^2 = M \qquad (31)$$

$$T = R_1 - W \qquad (32)$$

For any section between G′ and G″) we have:

$$R_1 x - W(x-m) - W_1(x-m_1) = \frac{1}{6} S b d^2 = M \qquad (33)$$

$$T = R_1 - W - W_1$$

For any section between G″ and C

$$\left.\begin{array}{l} R_1 x - W(x-m) - W_1(x-m_1) - W_2 \\ \quad (x-m_2) = \frac{1}{6} S b d^2 = M \\ \quad T = R_1 - W - W_1 - W_2 \end{array}\right\} \qquad (34)$$

The location of the greatest moment is most readily determined by geometrical construction.

Geometrically. The moments are represented in Fig. (30) by constructing

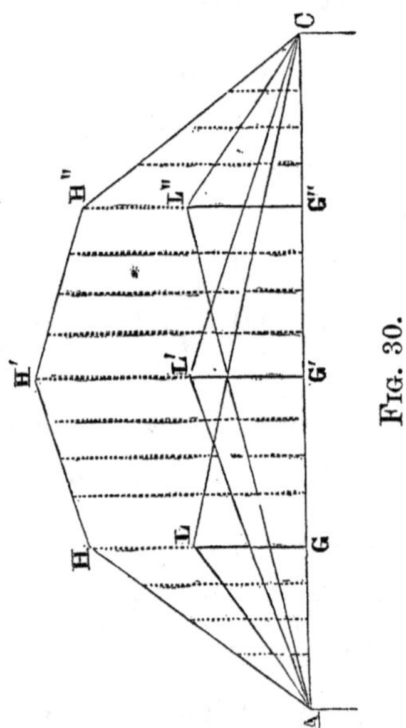

Fig. 30.

separately those due to each weight and then combining them. Thus, the moments produced at every point in the beam by the weight W are represented by the triangle A L C (Fig. 30), in which G L equals the greatest moment due to $W=\frac{n}{l} W m$. Similarly A L′ C represents those due to W_1, G′ L′ being$=\frac{n_1}{l} W_1 m_1$ and A L″ C gives those due to W_2, G″ L″ being$=\frac{n_2}{l} W_2 m_2$.

Now, if at every point we add together the ordinates of these three triangles for that point, and lay them off above A C we shall get a polygon A H H′ H″ C, which represents eqs. (29) (31) (33) and (34) and gives the total moment at any section. The greatest ordinate of this polygon will, of course, show the location of the maximum moment. This will be at G or G′ or G″, according to the relative amounts and positions of the weights W, W_1, and W_2. In the fig. it is at G′. Hence from eq. (31)

$$R_1 m_1 - W(m_1 - m) = \frac{1}{6} S b d^2 = M_0 \quad (35)$$

The shearing force may be represented

as in Fig. (31). It is greatest in that

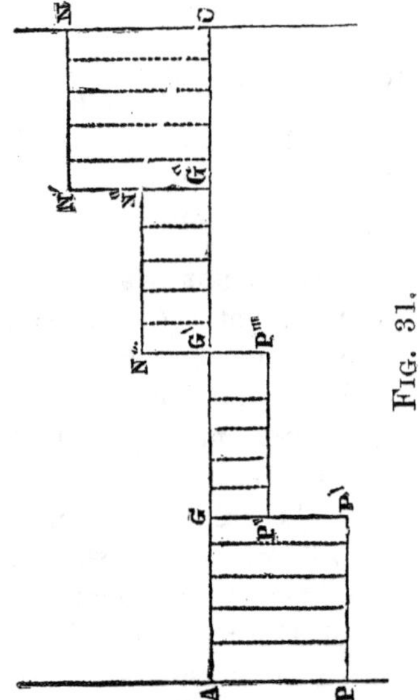

FIG. 31.

one of the two end segments which corresponds to the greater of the two quantities R_1 and R_2.

Note, That in this case the simplest way of finding the point of maximum moment is to construct the figure representing the shearing force, and the point when the shearing force passes through zero (G′ Fig. 31) is the point sought.

EXAMPLES.

1. Let l=20 ft. m=5 ft. m_1=10 ft. m_2=15 ft. W=1 ton W_1=2 tons W_2=3 tons. Find the maximum moment.

2. Find the size of a rectangular wooden beam where l=15 ft. m=3 ft. m_1=6 ft. m_2=14 ft. W=1 ton W_1=$\frac{1}{2}$ ton W_2=2 tons S=1,000 lbs. and d=4 b.

Case IV.

Let the load be equally distributed over the beam (Fig. 32). In this case the reaction of each abutment=$\frac{1}{2}$ the load, or

$$R_1=\frac{w\,l}{2}=R_2 \qquad (36)$$

Take any section E F whose distance from A=x. Then the external forces

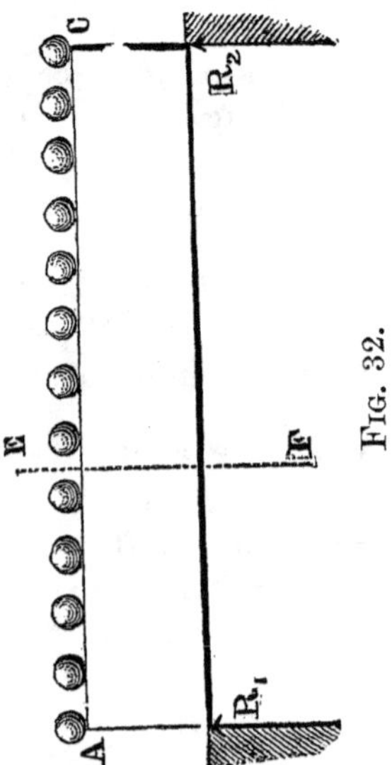

FIG. 32.

acting between A and E F are, R_1 and the resultant of all the little weights from A to E ($=w x$). This last force acts at its centre of gravity (Fig. 33), which is half way from A to E. Its lever arm is therefore $=\frac{x}{2}$. Hence the equation of moments will be

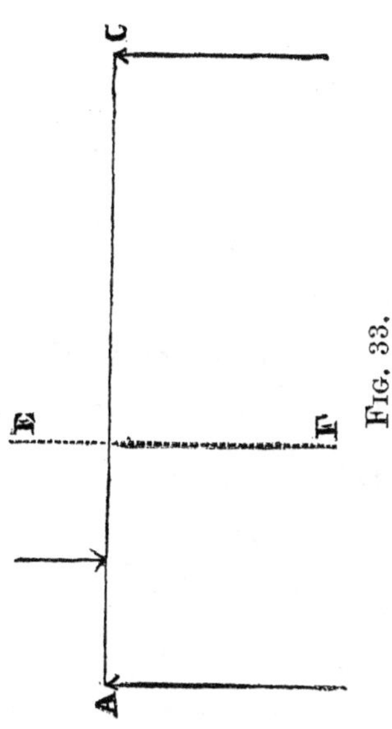

Fig. 33.

$$\left.\begin{array}{ll} & \frac{w\,l}{2} . x - \frac{w\,x^2}{2} = \frac{1}{6}\,S\,b\,d^2 = M \\ \text{or} & \frac{w\,x}{2}\,(l - x) = \frac{1}{6}\,S\,b\,d^2 = M \end{array}\right\} \quad (37)$$

This is a maximum at the centre where

$$M_0 = \tfrac{1}{8}\,w\,l^2 \quad (38)$$

The shearing force at E F is

$$T=\frac{w\,l}{2}-w\,x \qquad (39)$$

This is greatest at the abutments where $x=l$, or 0

$$\therefore\ T_0=\frac{w\,l}{2} \qquad (40).$$

At the centre $\qquad T=0$

Geometrically. The values of M in eq. (37) may be represented by a para-

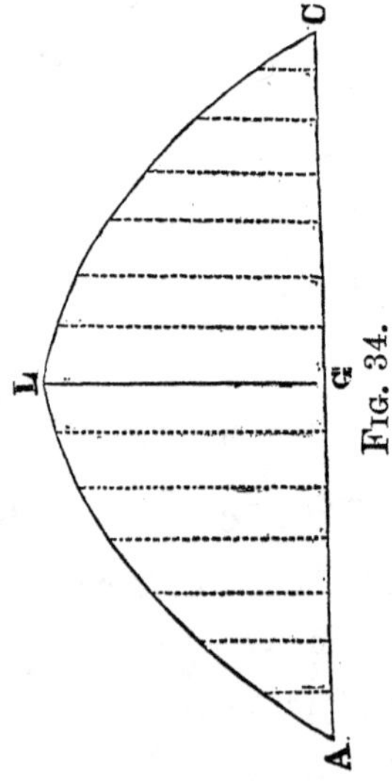

FIG. 34.

bola with vertex at L, the ordinates G L (Fig. 34) being taken $=M_0$. The shearing force is represented by the two triangles A P G and G Q C (Fig. 35).

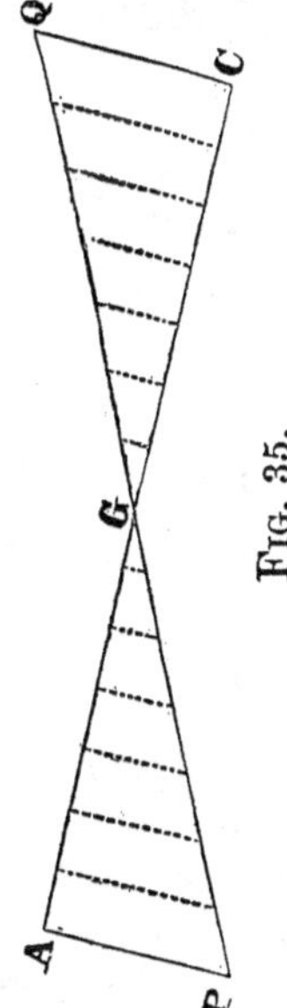

Fig. 35.

The maximum moment exists at the point (G) when the shearing force is zero.

Corollary. When the uniform load extends only over a certain distance from one of the supports, as in (Fig. 36),

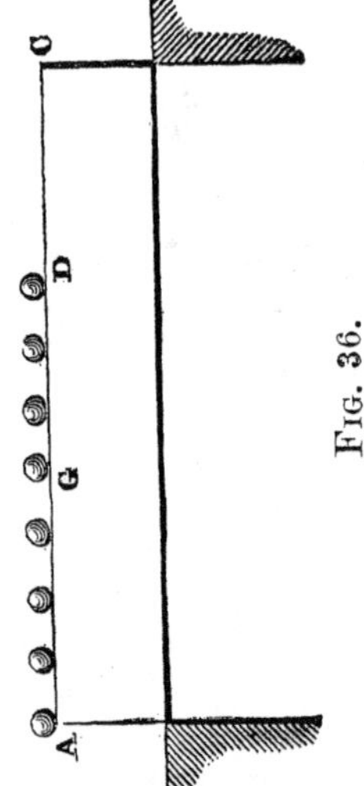

FIG. 36.

let A D=loaded segment=m. The reactions of the abutments are:

$$\left.\begin{array}{l} \text{At A, } R_1 = w\,m\left(\dfrac{l-\frac{1}{2}m}{l}\right) \\ \text{At C, } R_2 = w\,m\left(\dfrac{m}{2\,l}\right) \end{array}\right\} \qquad (41)$$

Then for any section in A D the moments of the external forces will be as in the case just discussed,

$$=w\,m\left(\frac{l-\frac{1}{2}m}{l}\right)x-\frac{w\,x^2}{2}$$

And the equation of moments in A D will be:

$$w\,m\left(\frac{l-\frac{1}{2}m}{l}\right)x-\frac{w\,x^2}{2}=\frac{1}{6}\,S\,b\,d^2=M \quad (42)$$

For any section in D C the whole load ($w\,m$) is to be considered as acting through its centre of gravity (G), and the equation of moments is:

$$wm\left(\frac{l-\frac{1}{2}m}{l}\right)x-wm(x-\tfrac{1}{2}m)=\frac{1}{6}S\,bd^2=M$$

Reducing

$$\frac{w\,m^2}{2\,l}\,(l-x)=\frac{1}{6}\,S\,b\,d^2=M \qquad (43)$$

The shearing force in A D is

$$\left.\begin{array}{l} T=w\,m.\left(\frac{l-\frac{1}{2}m}{l}\right)-w\,x \\ \text{In D C, } T=w\,m.\left(\frac{l-\frac{1}{2}m}{l}\right)-wm \end{array}\right\} \quad (44)$$

T is a maximum at A, or

$$T_0 = w\,m\left(\frac{l-\frac{1}{2}m}{l}\right)$$

Geometrically. Eq. (42) corresponds to the parabola A L K (Fig. 37), which

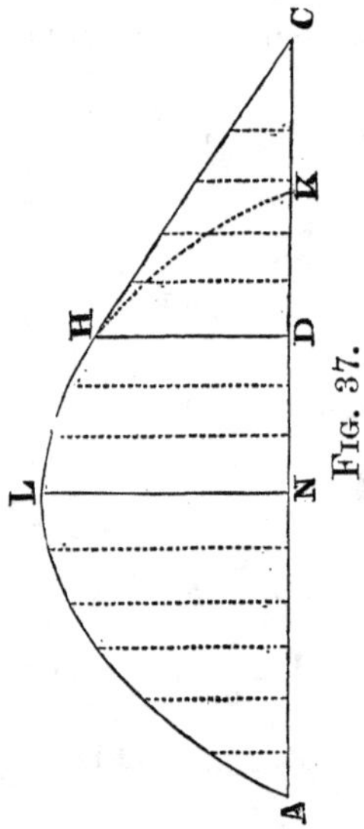

Fig. 37.

cuts A C at A and K (whose distance from A$=\frac{2m}{l}(l-\frac{1}{2}m)$) and whose axis is vertical. Eq. (43) corresponds to the straight line H C. We only use the part A L H of the parabola, the moments in D C being represented by the triangle D H C. The maximum moment is at N corresponding to the vertex L of the parabola. The value of this moment is:

$$M_0 = \frac{w\,m^2}{2}\left(\frac{l-\frac{1}{2}m}{l}\right)^2 \qquad (45)$$

which is obtained from eq. (42) by substituting for x the value A N $(=\frac{1}{2}$ A K$)=\frac{m}{l}(l-\frac{1}{2}m)$. This M_0 is always less than the maximum moment that exists when the load extends all over the beam as will appear by making m to vary in eq. (45) and applying the tests for a maximum to it.

The shearing stress for the loaded segment is represented by the triangles A P K and K P′ D (Fig. 38), and for the other segment by the rectangle D H. The point K, where the stress is zero, is found by making in eq. (44)

$$T = wm\left(\frac{l - \frac{1}{2}m}{l}\right) - wx = 0$$

and finding the value of x.

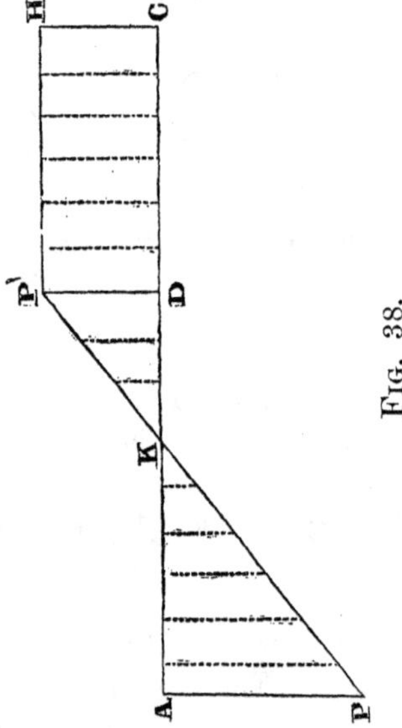

FIG. 38.

This point corresponds to the maximum moment. It may also be found graphically by constructing Fig. 38, and, as before, affords the easiest method of determining the point of the beam where the maximum moment exists and where consequently there is greatest danger of rupture.

EXAMPLES.

1. Let $l=20$ ft. $w=500$ lbs. per ft. $m=15$ ft. and let there be a weight in

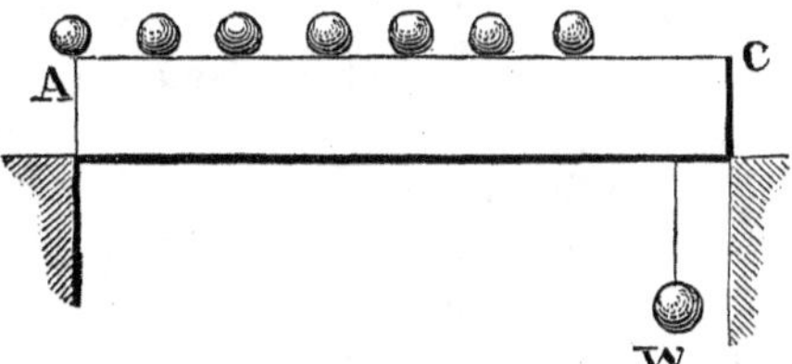

Fig. 39.

addition. W=5 tons at a point 18 ft. distant from A. Required the maximum moment.

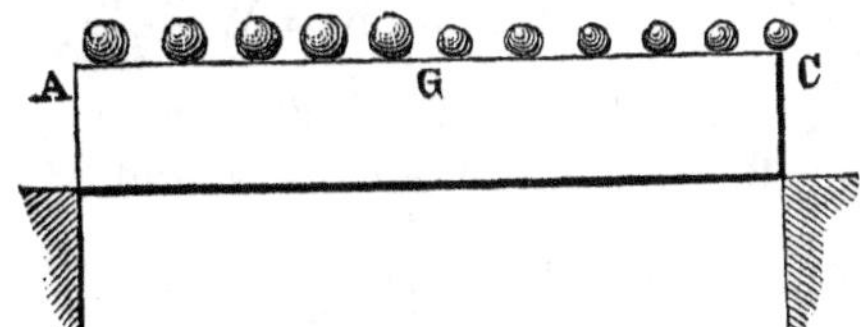

Fig. 40.

2. Let one-half of the above beam be loaded with a uniform load, $w=1$ ton per foot, and the other half with a uniform load of $w'=\frac{1}{2}$ ton per foot. Required the moments.

Case V.

A single moving load. When a single moving load passes over a beam, as in (Fig. 41), the maximum moment at each instant (as appears from *Case III.*) takes effect at the section just under the weight. To determine the law of variation of these maxima as the weight travels over the beam: Let x=the distance at any instant from A, and then the reaction of A at that instant (=the part of W transmitted to it)$=\text{W}\,\frac{l-x}{l}$ Multiplying this by the lever arm x we have for the moment under the weight:

$$\frac{\text{W}\,x}{l}\,(l-x)=\frac{1}{6}\,\text{S}\,b\,d^2=\text{M} \qquad (46)$$

This is a maximum at the centre, where

$$M_0 = \tfrac{1}{4} W l \qquad (47)$$

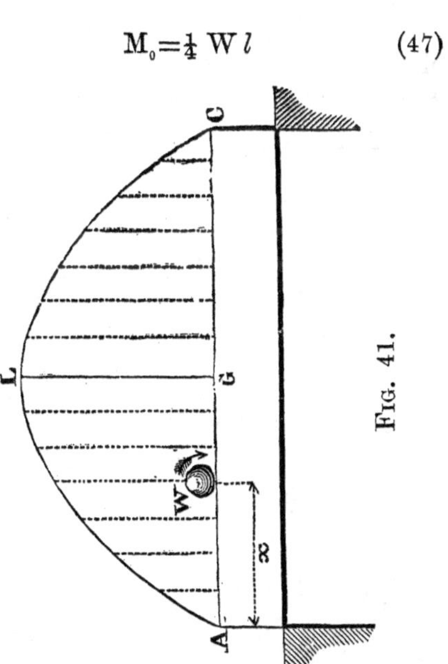

Fig. 41.

Eq. (46) corresponds to a parabola (Fig. 41) with vertex at L, the ordinate G L being $= \frac{1}{4} W l$. The shearing force for each segment into which W (Fig. 42) at any instant divides the beam is equal to the reaction of the abutment corresponding to that segment. Thus, if W is at a distance x from A the reaction of A is $=W\frac{l-x}{l}$ and of C it is $= W\frac{x}{l}$.

If W has the position marked 2 in (Fig. 42), then the shearing stress in the

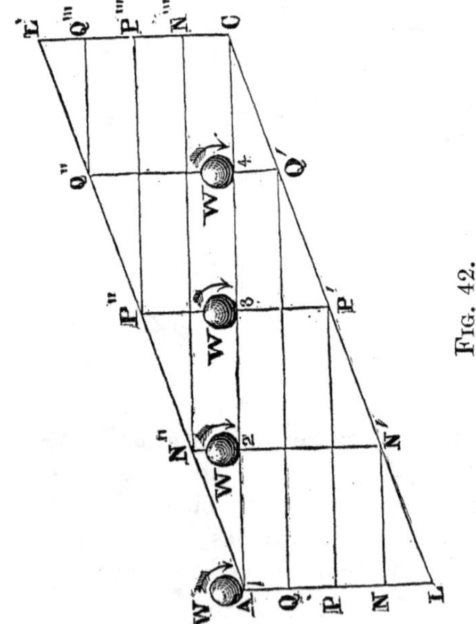

Fig. 42.

left segment is shown by the rectangle A N N′ W and in the right segment by the rectangle W N″ N‴ C. The diagram shows in a similar way the stress at other points. If the third position of W in the figure is at the centre of the beam then evidently the greatest shearing stress to be provided for in the left half of the beam will be represented by the *locus* of the points like L, N′, P′, and for the right half it will be the *locus* of the points P″, Q″, L′, etc.

These *loci* are given by the equations:

$$\left.\begin{aligned} T&=\frac{W}{l}(l-x)=\text{eq. of L C} \\ T&=\frac{W}{l}\,x \qquad =\text{eq. of A L}' \end{aligned}\right\} \quad (48)$$

Turn the triangle A L′ C down for convenience, as in (Fig. 43), and then the shearing stress to be provided for is given by the figure A L P L′ C. In this figure A L = C L′ = W and D P $=\frac{W}{2}$.

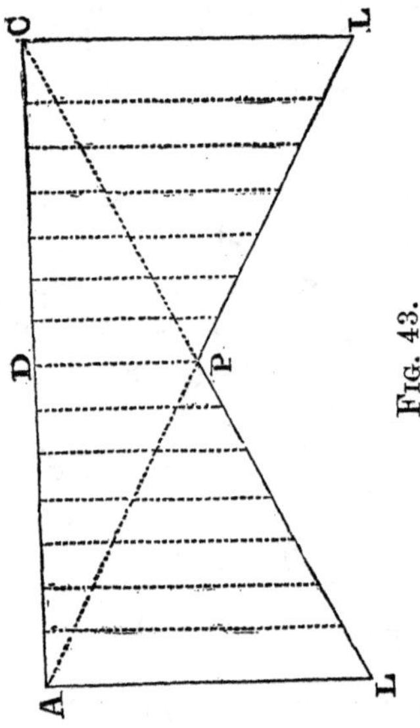

FIG. 43.

Case VI.

A distributed moving load. When a moving load gradually covers a beam, (Fig. 44), moving on from one end as a long train of cars, the maximum moments produced is that due to the load when it covers the entire length of the

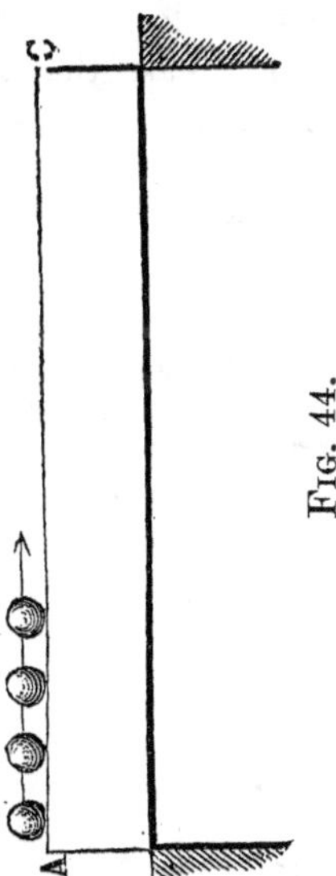

Fig. 44.

beam, and consequently this case is provided for in *Case IV.*

But with the shearing stress it is different. Here, as in *Case V.*, we need the *locus* of the greatest shearing stresses that can be brought upon the beam. This maximum at any section D occurs when the longer segment into which D divides the beam is loaded, and the other is not. In that case the shearing force at D (=the reaction of the abutment C) is

$$T=\frac{w\,x^2}{2\,l} \qquad (49)$$

This equation gives the parabola A N P′ (Fig. 45) with vertex at A, where

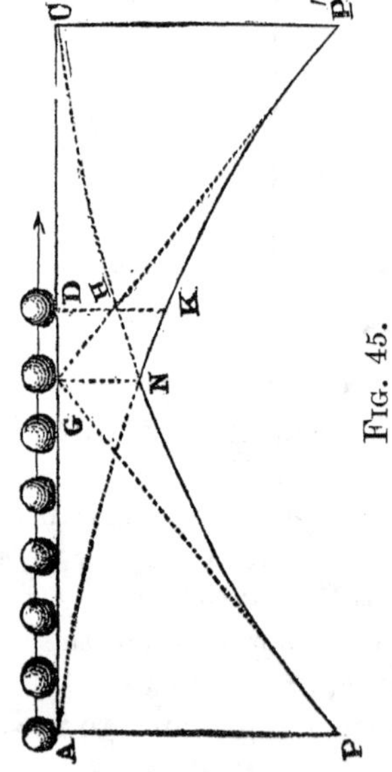

Fig. 45.

$CP' = \frac{wl}{2}$. When the load comes from the other end of the beam we get the parabola C N P. Hence the figure A P N P′ C gives the maximum shearing stress to be provided for.

It is easy to see that the shearing stresses thus obtained are greater than those which exist when the load covers the entire beam. In the latter case the forces are represented by the triangles A P G and G P′ C (Fig. 45), the shearing stress at D being given by eq. (39)

$$T = wx - \frac{wl}{2} = DH$$

In the case of the passing load we have just seen that

$$T = \frac{wx^2}{2l} = DK$$

The value of D H is always less than that of D K when $x \geq \frac{l}{2}$; for if $2x$ be a certain quantity, then the product of the halves of that quantity $(= x^2)$ is greater than the product of any other two parts (such as l and $[2x-l]$) into which it can be separated.

That is

$$\left.\begin{array}{ll} & x^2 > l\,(2x-l) \\ \therefore & \dfrac{w\,x^2}{2\,l} > \dfrac{w\,l\,(2x-l)}{2\,l} \\ \text{or} & \dfrac{w\,x^2}{2\,l} > w\,x - \dfrac{w\,l}{2} \end{array}\right\} \quad (50)$$

In the expressions for the moment of resistance $M = \frac{1}{6}\,S\,b\,d^2$ the quantity denoted by S (= the stress on the outside fibres) varies directly as M. Hence, all the geometrical illustrations we have given of the *moments* may apply equally well to the values of S. The maximum moments give the maximum stress on the fibres, and indicate the points of rupture when the beam is loaded with its breaking weight.

ULTIMATE VALUES OF S.

If beams are loaded transversely until fracture takes place, the value of S or the stress on the outside fibre which exists at the moment of fracture, gives us a value for the tensile or compressive strength of the material according to the manner of rupture. If the beam yields by *tearing*, S gives us the tensile strength, if by *crushing* S gives the compressive strength. We readily obtain the value of S answering to the ultimate strength from any of the formulas under "Transverse Stress," by substituting given values for *l*, *b*, and *d* and the actual breaking weight for W.

But the tensile and 'compressive strengths of materials are also obtained by direct tension and compression, the force being applied in the direction of the length of the bars until rupture takes place.

If our theory were perfect the values of tensile and compressive strength thus deduced would agree with the ultimate values of S found in transverse stress; but they do not.

The difference is very wide sometimes. Thus in cast-iron, S (in this case it represents the tensile strength) derived from breaking rectangular beams by a transverse load is nearly 20 tons per square inch, while the tensile strength obtained directly is only about 8 tons. This discrepancy has been accounted for in two ways.

1. That the neutral axis moves towards the compressed side, and that therefore a larger portion of the beam is subjected to tension than the formula supposes.

2. That the neutral axis always remaining at the centre of gravity of the beam, the additional strength is due to the *adhesion* of the fibres which is developed by the unequal lengthening, and shortening of them as we go from the neutral axis towards the surfaces. In favor of this view is the fact that we know such adhesion to be an element of strength; for the compression or extension due to a given force is not so great in a transversely loaded beam as in one directly compressed or extended.

The action of this adhesive force may be illustrated as follows :

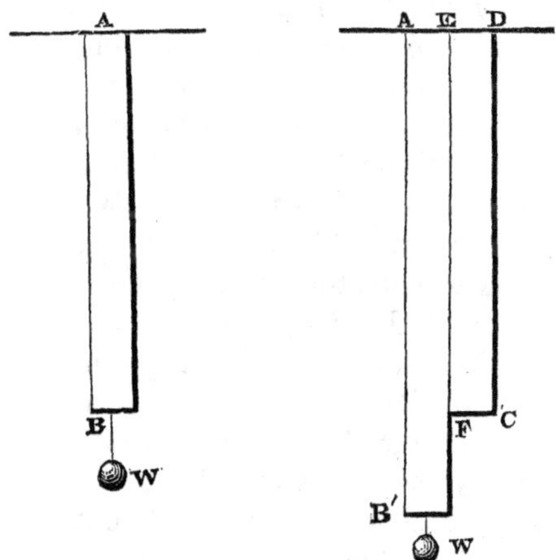

Fig. 46.

In the beam AB (Fig. 46), strained by the weight W, all the fibres are equally elongated, and they only resist by their direct tenacity. But in the beam A′C to the one-half of which is appended the weight, while the other half, EC, is less strained or altogether prevented from extending, evidently W will have to overcome not merely the tenacity of the fibres in A′B′ but the adhesive force of the fibres along the plane E F, where the two parts of the beam join ; for this force will tend to prevent the stretching of the fibres in A′ B′, and consequently increases the strength of A′ B′. This kind of force exists between every two layers of horizontal fibres in a beam under transverse loading, and is called the *longitudinal* shearing stress. It is neglected in the formulæ we have given.

From the variation between the ultimate valves of S (called *moduli of rupture*) and the values for strength obtained by direct tension and compression, it results that the values should be determined in bc th ways, and that the values gotten by one method should not be used in calculations involving the other kind of stress.

BEAMS OF UNIFORM STRENGTH.

As already stated, in solid rectangular beams, S has different values for the various points in the length of the beam. There is always a point of maximum stress where the beam, if loaded sufficiently, will break. Now at all other points there is an excess of material which is useless and injurious from its own weight. To secure the requisite strength with the least material is an object usually desirable, and this can be readily accomplished in certain materials (as cast-iron), by giving the beam such a shape as will make S, the stress on the outside fibre, constant throughout its length. In wood the injury resulting from the cross cutting of the fibres frequently prevents the putting of the theory into practice.

The application of the theory of uniform strength to beams of *rectangular cross section* may be most simply explained by taking up the cases we have discussed in detail.

In *Case I.* from eq. (5) the maximum stress in the outside fibre is,

$$S = \frac{6\,W\,l}{b\,d^2} \qquad (51)$$

This stress only occurs at A, where the

beam will ultimately break, and it is evidently possible to take away some of the material between that point and

FIG. 47.

C without diminishing the strength. If this be so done that at every point between A and C there shall exist on the outside fibre a stress equal to that at A, the beam will be one of uniform strength, and we shall have attained the greatest economy of material. Let us suppose, the use we have for the beam requires the depth to be uniform. What must be its *plan* in order that S shall be constant in value, or the beam be as liable to break at any other point as at A^5.

In eq. (4) $S=\frac{6\,W\,x}{b\,d^2}$, if we assume S to

be constant, the other side of the equation must be constant also, and since $6W$ is constant, and we have made d constant by assuming the depth to be uniform, the whole expression can be constant only when b varies as x.

$\therefore$ $b \propto x$ whence $b = cx$ (where c = some constant factor). This equation which is that of a straight line shows that the breadth must vary directly as the length. Hence the *plan* should be a triangle with vertex at C (Fig. 48.)

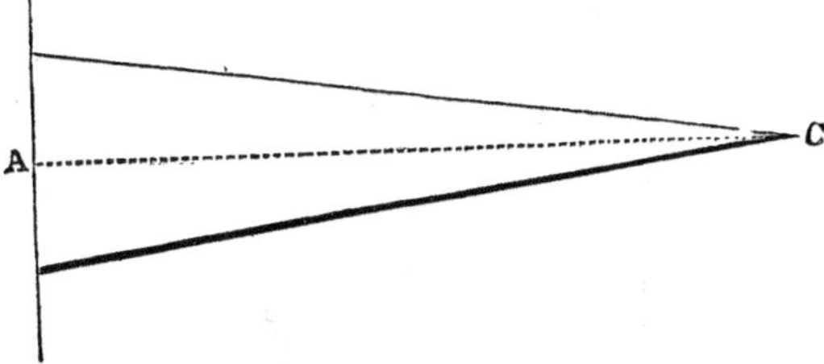

FIG. 48.

On the other hand, if we suppose the breadth to be uniform and wish to have S constant, in the eq. $S = \frac{6Wx}{bd^2}$, x must vary as d^2, or

$$d^2 = cx$$

This corresponds to a parabola, and the beam, if the top be straight will have the *elevation* shown in (Fig. 49).

Suppose that b varies as d, then $d=nb$ (n being a constant) and

$$S=\frac{6\,W\,x}{n^2\,b^3}$$

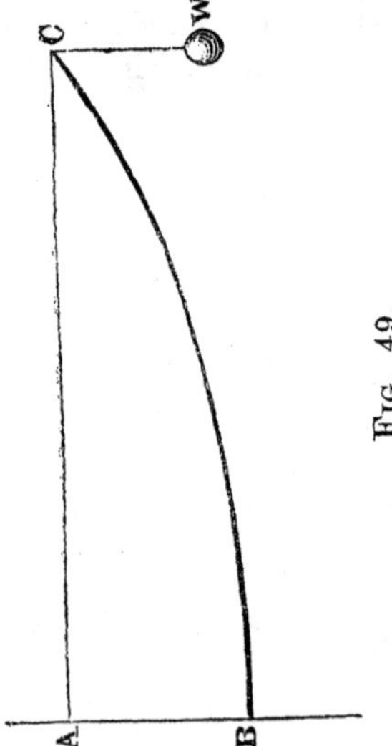

Fig. 49.

To render S. constant we must have, $b^3 \propto x \;\therefore\; b^3 = c\,x$ and $d^3 = n^3\,c\,x$.

These are the equations of a *cubic parabola.* Hence the horizontal section (Fig.

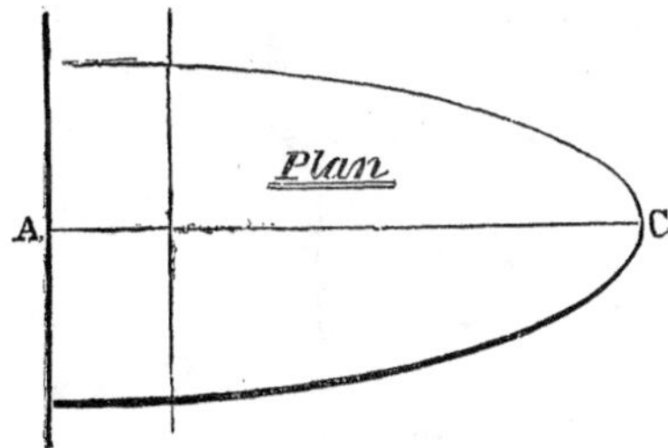

Fig. 50.

50), and the vertical section (Fig. 51), should be curves of that kind. The

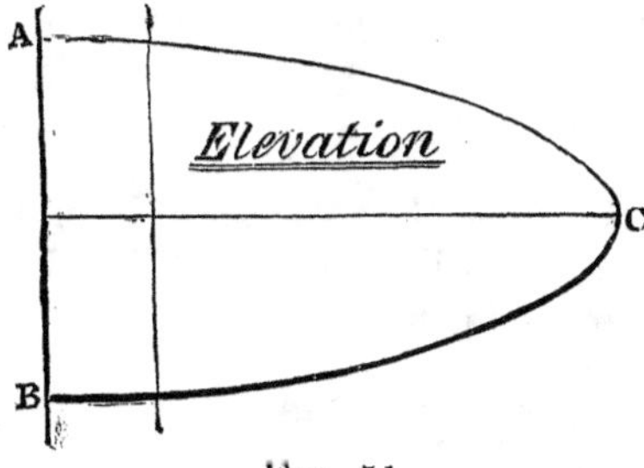

Fig. 51.

cross section is rectangular as in (Fig. 2).

In *Case II.* we have

$$S=\frac{3\,W\,x^2}{b\,d^2} \qquad (55)$$

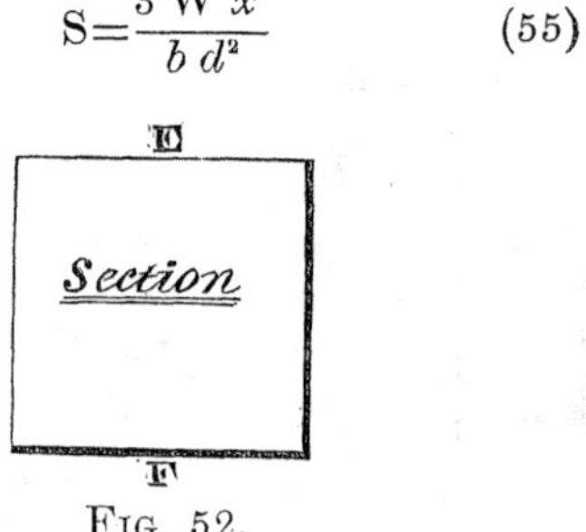

Fig. 52.

Hence, if we make suppositions similar to those above we shall have, when the depth is uniform (or $d=$ a constant) x^2 varying as b, or

$$b=c\,x^2$$

Hence the *plan* (Fig. 53) should consist of parabolas with vertices at C. If b be constant then

$$d^2=c\,x^2 \quad \text{or} \quad d=\sqrt{c}.\,x$$

This is the equation of a straight line,

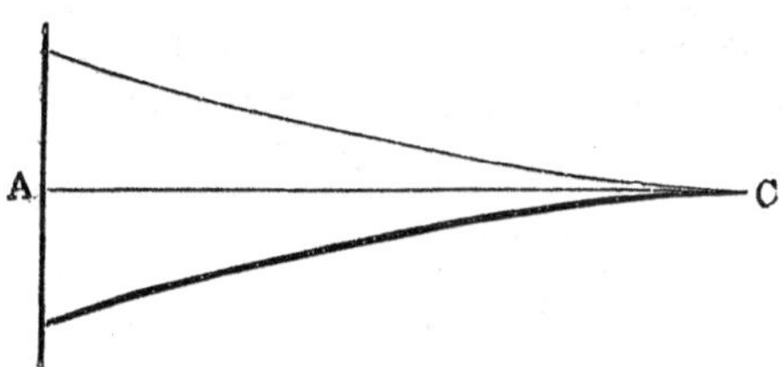

FIG. 53.

and gives for the *elevation* the triangle (Fig. 54).

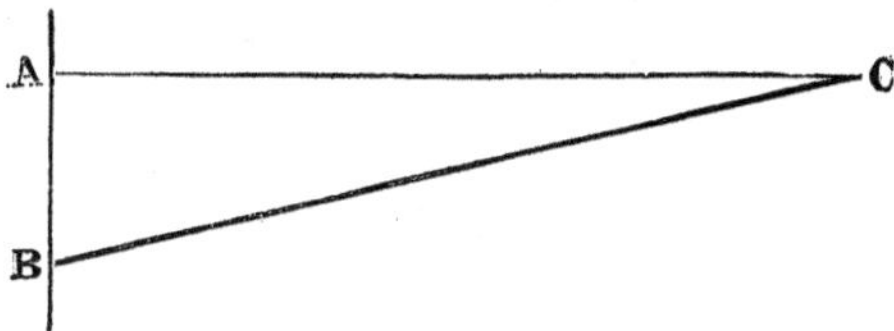

FIG. 54.

In *Case III.* the analysis gives results similar to those in *Case I.*

Thus from equation (21) $S=\frac{6\,W\,n}{l}\cdot\frac{x}{b\,d^2}$. Here $\frac{6\,W\,n}{l}$ is constant, and if d be constant also, b must vary as x.

$$\therefore\quad b=c\,x.$$

This gives the triangle A G H (Fig. 55). We obtain similarly the triangle G C H for the other end of the beam.

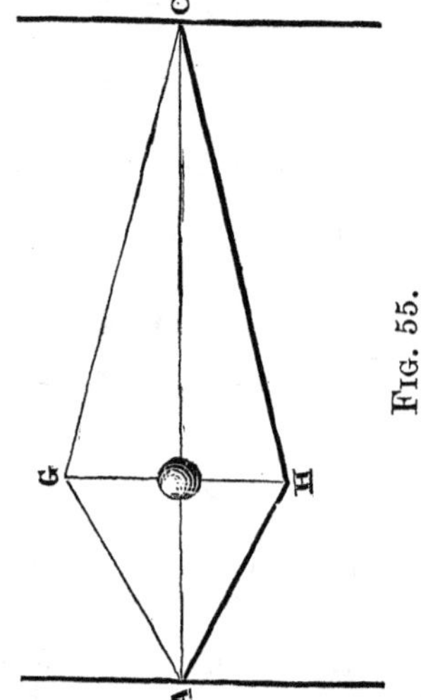

Fig. 55.

If the breadth be constant we have,

$$d^2 = cx$$

which gives a parabola AK (Fig. 56) with vertex at A. So, for the right hand end of the beam the proper elevation is the parabola C K (Fig. 56). The elevation (Fig. 56) assumes that the top of the beam needs to be horizontal.

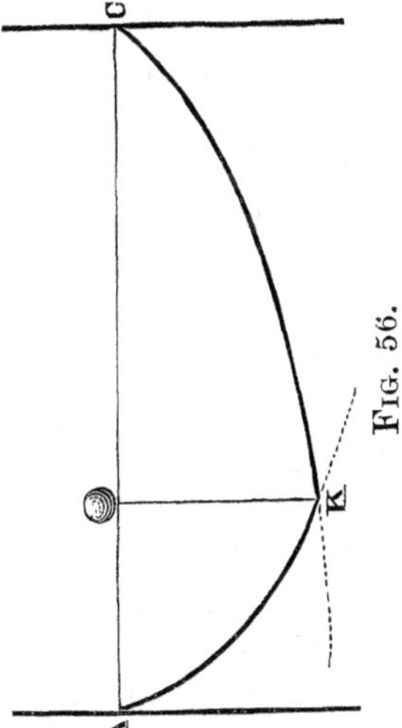

FIG. 56.

In *Case IV.*, equation (37) gives $S = \frac{3\,w\,x\,(l-x)}{b\,d^2}$: Here, if d be constant in order to render S constant we have

$$b = c\,x\,(l-x).$$

This may be represented by parabolas with vertices at G and H (Fig. 57) opposite the middle of the beam. If b is constant, then,

$$d^2 = c\,x\,(l - x)$$

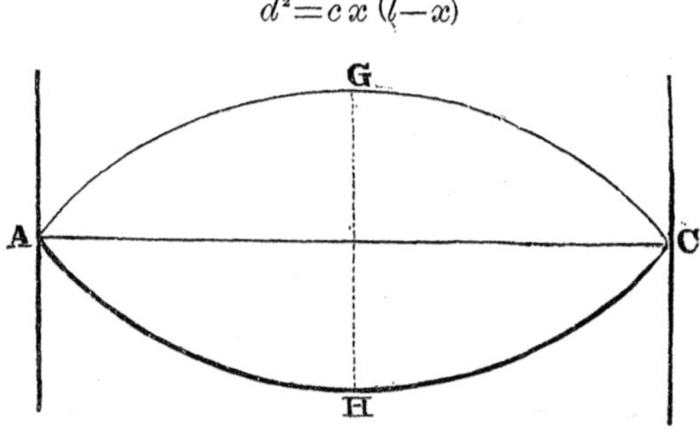

Fig. 57.

which is the equation of an ellipse, and the beam (if it is required to be horizontal on top) may be made as in (Fig. 58.)

In these cases of beams of uniform strength, we have so far only considered the *moments* of the weights or the *bending moments* as they are called. But the results are to be modified by the transverse shearing stress. In ordinary rec-

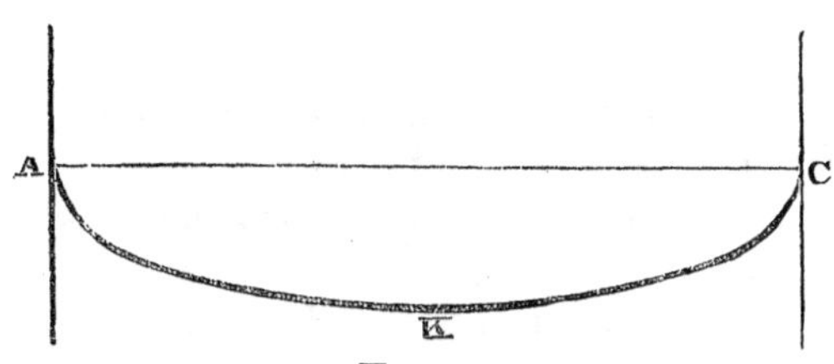

FIG. 58.

tangular beams this shearing stress is so small compared with the bending moment, that it may be left out of consideration. But in beams of uniform strength the ends must not taper to a point, but must always be left large enough to bear the shearing stress. In the case represented in (Fig. 57) the beam should have, near the ends, the shape shown in (Fig. 59).

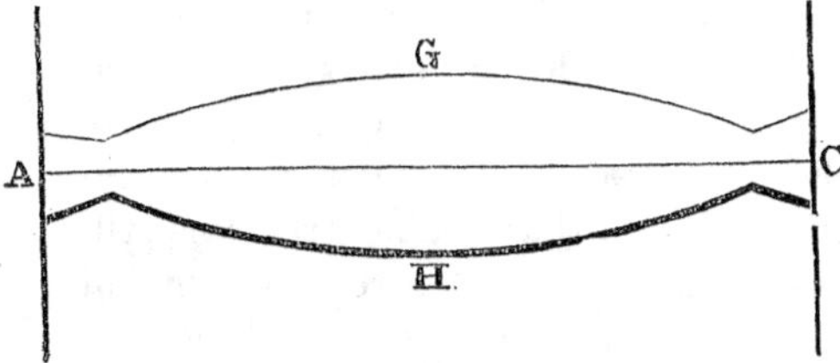

FIG. 59.

DOUBLE FLANGED BEAMS.

So far we have considered beams with rectangular cross sections, and beams of uniform strength deduced from these. We will now consider beams of ⌶ shape.

It is evident from the investigation already given of the condition of stress in transversely loaded beams, that those portions of the beam nearest the centre bear but a small proportion of the stress, while the contrary is the case with the outside fibres. Hence we would gain strength by moving a considerable portion of that about the neutral axis and placing it on the top and bottom.

The first form in which the idea was applied was in the **T** or **⊥** cast iron beam. The fact that rectangular cast-iron beam always broke by the tearing of the fibres on the side subjected to tension, suggested the idea of reinforcing that side of the beam with a flange. The result of this is, that the neutral axis still passing through the centre of gravity of the cross section, the extreme fibres subjected to compression are farther off than those subjected to tension, and consequently are strained more nearly to their full strength before fracture. This form of beam gives a large increase of strength for the same amount of iron.

It was still plain that the fibres in that part of the web about the neutral axis were but little strained as compared with the fibres on the outside, and it was proposed to leave as little material there as possible, and to place the mass of it in two flanges (⌶), one above and the other below, giving to these flanges sizes inversely proportional to the tensile and compressive strength of the material. The question then was, how much of the material should be left in the *web*, for plainly all could not be taken. The amount to be left is determined by experiment. If the web is left too thin, the beam will twist and break under the shearing force, and in some cases, from the want of stiffness in the compressed flange.

To simplify the calculations, the web

is considered as bearing all the shearing stress, and no other, and the flanges as bearing all the extension and compression due to the bending moment ; and these parts should be proportional accordingly with due reference to the practical difficulties that sometimes occur. The ordinary formulas for the strength of such beams are gotten by the following approximation: We first neglect the compressive and tensile forces of the web, which are small compared with those of the flanges, and consider it as bearing only the shearing stress. Then as the depth of the flanges is generally small as compared with the depth of the beam, we consider all the fibres in each flange as strained alike, and as bearing the average stress that is brought on that flange. (Fig. 60.)

The resultant of the force on each flange, then, is equal to the stress on a unit of surface (S) multiplied by the flange area (A) : that is = S A.

The point of application of the force will be at the middle of the depth of the

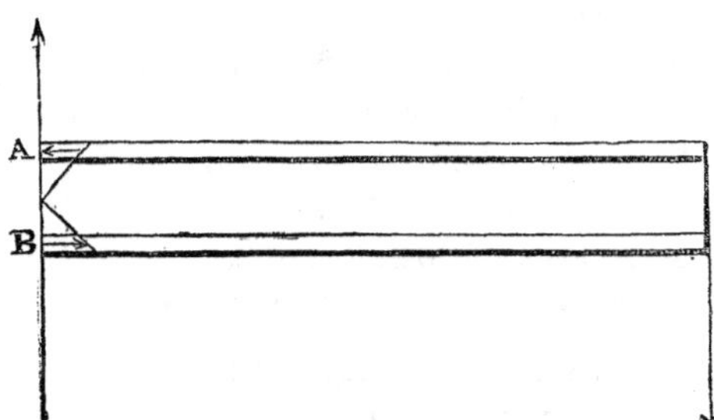

Fig. 60.

flanges (at O and O′, Fig. 61). Fig. 61

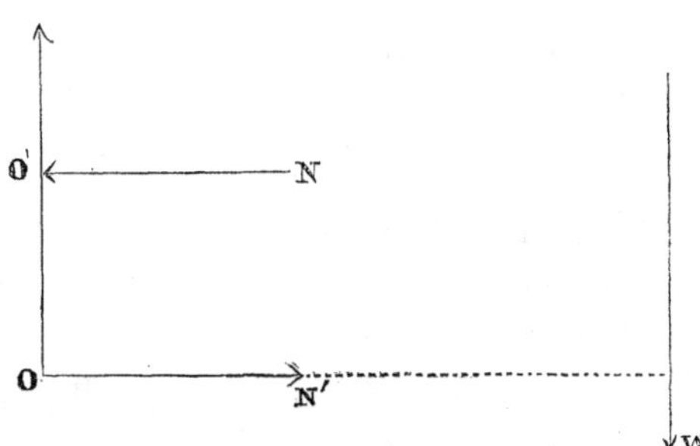

Fig. 61.

shows the forces we have to deal with in the *Case* corresponding to *Case I.* under rectangular beams.

Let $O' O = d$.

$S' =$ stress on upper flange per unit of surface.

$S'' =$ stress on lower flange per unit of surface.

$A'' =$ area of lower flange.

$A' =$ area of upper flange.

Then if we take O (Fig. 61) as a centre of moments we have :

$$-N d - N'.0. + W x = 0 \quad (\text{But } N = S' A')$$

$$\therefore S' A' d = W x \qquad (57)$$

If we take O′ as the centre of moments we will get

$$S'' A'' d = W x \qquad (58)$$

The formula for shearing force is identical with that under *Case I.* of rectangular beams ; that is :

$$T = W x \qquad (59)$$

If $A' = A''$, then plainly $S' = S''$ (from equations 57 and 58), or, the forces of tension and compression are equal (as in rectangular beams) ; but if A' and A'' are not equal, we have :

$$S' : S'' :: \frac{W x}{A' d} : \frac{W x}{A'' d} :: A'' : A'$$

That is, the unit stresses in the flanges are inversely as the areas. Now, to have the material distributed between the flanges most efficiently for strength the unit stress should be in proportion to the ultimate strength of the material against tension and compression, and hence the areas of the cross sections of the flanges should be *inversely* as the ultimate strength.

Thus, if A D (Fig. 62) be of cast-iron, which is six times as strong against compression as against tension, the unit stress in the lower flange should be made six times as great as in the upper, and to effect this the area of the lower flange should be one-sixth that of the upper.

The *Cases* under ⌶ beams are similar to those under rectangular beams.

Cass I.—Beams fixed at one end and loaded at the other.

$S' A' d = W x$, and $S'' A'' d - W x$ (60)

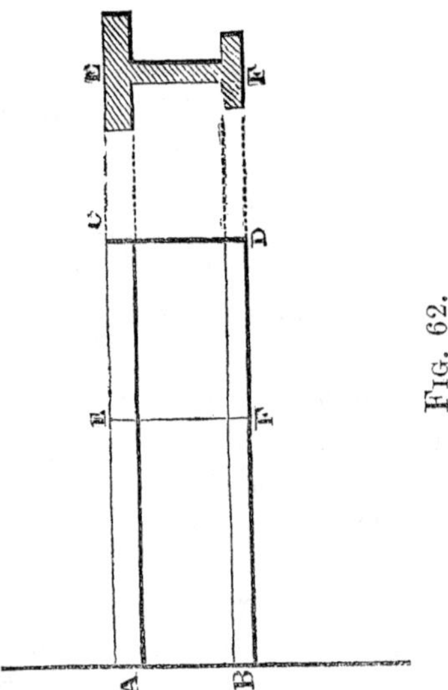

Fig. 62.

Case II.—Beams fixed at one end and loaded uniformly.

$$S' \, A' \, d = \tfrac{1}{2} \, w \, x^2, \text{ and } S'' \, A'' \, d = \tfrac{1}{2} \, w \, x^2 \quad (61)$$

Case III.—Beams supported at both ends and loaded at some intermediate point.

$$W.\frac{n}{l}.x - W(x-m) = S' A' d, \text{ or,} = S'' A'' d \quad (62)$$

Case IV.—Beams supported at both ends and loaded uniformly.

$$S' A' d = \tfrac{1}{2} w x (l-x) = S'' A'' d \quad (63)$$

Case V.—A single moving load over a beam supported at both ends.

$$S' A' d = \frac{W x}{l} (l-x) = S'' A'' d \quad (64)$$

Case VI.—A distributed moving load may be considered as included in *Case IV.*

The formulae for shearing stress are identical with those in rectangular beams.

The principles of the *uniform strength of beams* may be applied to flanged beams as they were to rectangular beams. The discussion is analogous to that already given.

MOMENT OF RESISTANCE OF BEAMS DETERMINED GEOMETRICALLY.

The following method of obtaining the moment of resistance of beams is of easy application, and in many cases of unsymmetrical cross section is the simplest that can be used :

I. For illustration, take a beam of rectangular cross section. Let G P (Fig. 63) be the cross section at some point of this beam. The stresses on the fibres, as we have already seen, increase just in proportion as we go from the neutral axis towards the upper or lower surface of the beam, and may for any vertical slice (as that at E F) be represented by the ordinates of two triangles, as shown in Fig. 64, where E I (=F J) represents the stress on the outside fibre. For the cross section G P (Fig. 63) the stresses will be represented by two wedges, the bases of which are G M and M R, and the elevations of which are the triangles shown in Fig. 64. The volumes of these wedges give the amount of compressive

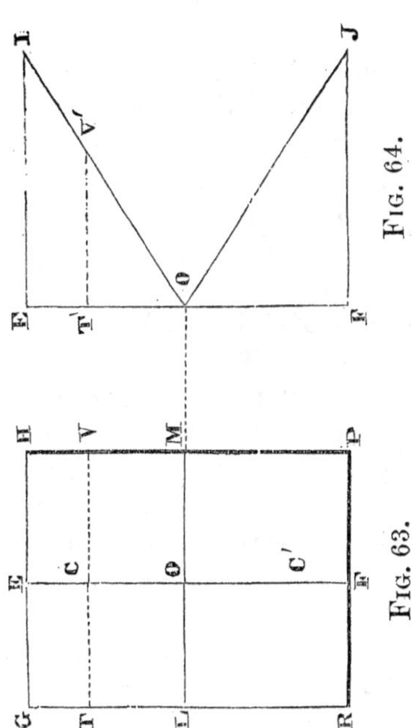

FIG. 63.

FIG. 64.

and tensile force exerted at the cross section in question, and the points in GP under the centre of gravity of the wedges give the "centres of resistance," or the points of application of the resultants of these forces.

As a geometrical representation of the stresses on the fibres, these wedges are perfect, for the perpendicular ordinate of the wedge gives in every case the stress which exists in the fibre over which it stands. Thus the line T′ V′ (Fig. 64) represents the stress on each fibre in the row T V (Fig. 63).

But it is often difficult to find the centre of gravity of these wedges in the case of curved and irregular cross sections, and yet this must be done before we can know the lever-arms of the stresses. To render this easier to do we may represent the stresses, not by wedges, but by *prisms*, the centres of gravity of which are over the centres of gravity of their bases.

Thus, if in (Fig. 65) we draw the two shaded triangles, and conceive prisms of a height = E I (the stress on the outside fibre, Fig. 64), to be constructed on them as bases, we shall have a geometrical representation of the stress on the section G P, less perfect in some respects

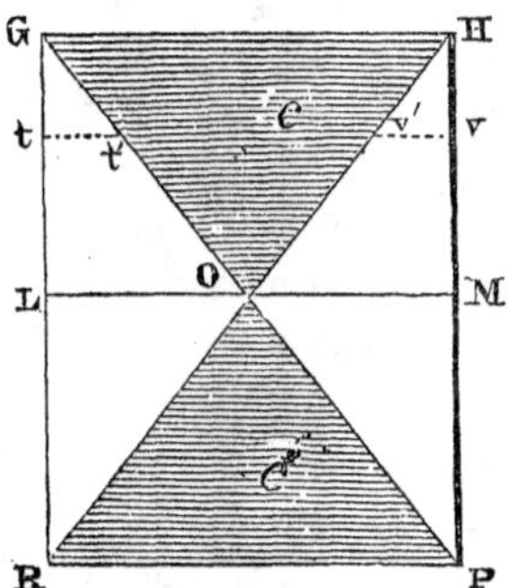

FIG. 65.

than that given by the wedges but better suited to our purpose.

For, note that,

1. The volume of the prism G O H (Fig. 65) is equal to that of the wedge G M (Fig. 63), and the volume of any part of the prism cut off by a plane parallel to the neutral axis, as that whose base $t' O v'$ is equal in volume to the corresponding part of the wedge T M, or since the height of the prism is constant, the stress on the surface G M as we go out from the neutral axis varies

as the area of the triangle which forms the base of the prism.

2. The vertical slice of the prism standing on any line $t' v'$ represents in amount the stress on the line of fibres $t v$, for this stress is equal to the corresponding one in the wedge, the slice of the prism being as much higher than that of the wedge as $t v$ exceeds $t' v'$. Of course (except in the case of the outside fibres in the row G H) each ordinate in the slice of the prism no longer represents the stress on the fibre over which it stands, as was the case in the wedge.

3. The moment of the tensile forces, for instance, will equal the area of the prism G O H multiplied by its height, (E I = stress on outside fibre = S). The centre of resistance of these forces, or the centre of gravity of the prism is at C (Fig. 65), the centre of gravity of the base G O H. The triangle G O H is sometimes called the "effective area" of the surface G M, because a uniform stress on it of an intensity = the unit stress at G H gives the same amount of

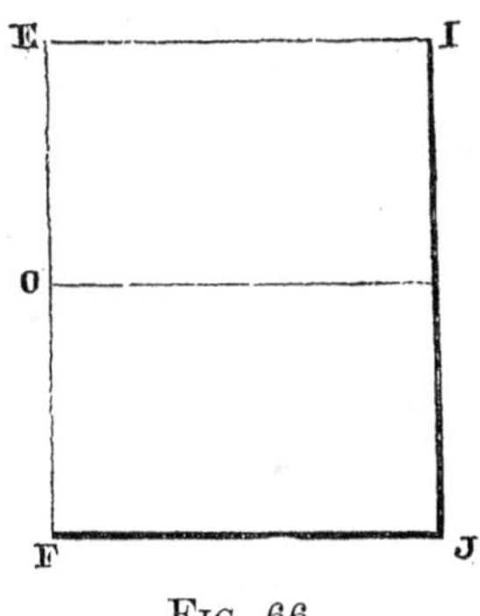

FIG. 66.

resistance, as that on the whole area G M, acted on as the latter is by a varying stress.

Considering the stresses represented by the two prisms whose bases are G O H and R O P (Fig. 65), as concentrated at the centres of gravity C and C′ (Fig. 67) of these bases, and taking one of these points (as C′, Fig. 67) as the centre of moments, we have in the case represented in the figure :

$$(\text{Vol. of prism } G\,O\,H) \times C\,C' = W\,x$$

or if $b =$ breadth and $d =$ depth of beam

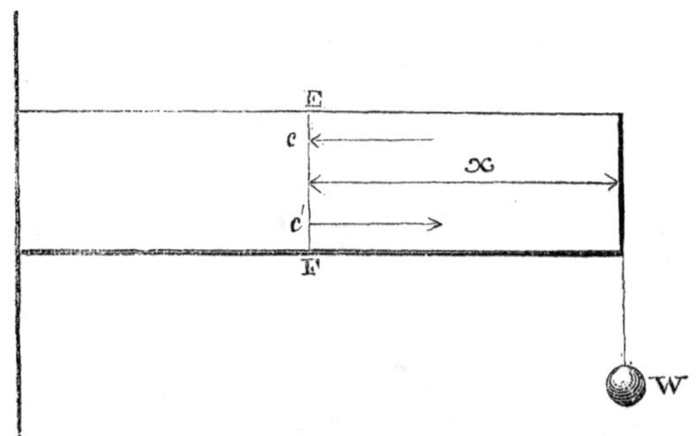

Fig. 67.

$$S\left(\tfrac{1}{4}\,bd\right).\,\tfrac{2}{3}\,d=\frac{1}{6}\,S\,b\,d^2=M=W\,x \quad (65)$$

as before.

Corollary. If the beam be square, $b=d$, and

$$M=\frac{1}{6}\,S\,b^3 \quad (66)$$

II. As a second example, take a square beam so placed that its diagonal will be vertical. Fig. 68 is the cross section. Here we find the base of the prism of stress by points. To find the line in the base of the prism corresponding to the

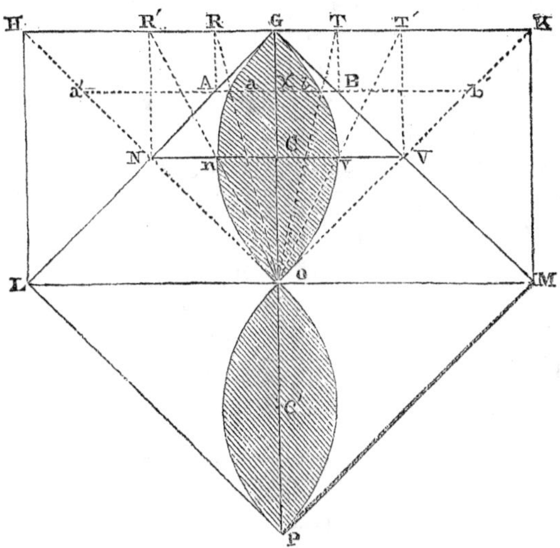

FIG. 68.

stress in any row of fibres, such as A B whose distance from the neutral axis is O X, proceed as follows :

We see that if the cross section were the square of which H L M K is the half, then a' b' would be the line required, since this is the breadth at that point of the triangle H O K, which would in that case represent the base of the prism of stress. Project the points A and B upon H K. Then the actual row (A B) of fibres is as much shorter than the corresponding row in the supposed section H M, as R T is less than H K, and consequently to obtain the proper line in the base of the true prism of stress, $a' b'$ must be shortened in this proportion. Draw lines from R and T to O. These lines intersect the row of fibres at a and b. Then

$$HK : RT\ (=AB) :: a' b' : ab \qquad (67)$$

Hence $a b$ is the line required, and a and b are two points in the outline of the base of the prism of stress. Any number of lines as $n v$, &c., may be gotten similarly and the curve drawn through the points $a \cdot\cdot n \cdot\cdot b \cdot\cdot v$, &c., will give the form of the base of the prism of stress. This base is shaded in the diagram.

For any ordinate of the curve O n G, as a X, we have

$$OX : aX :: OG : RG = AX$$

But $AX = XG$ and making $GO=\frac{d'}{2}$ and putting $OX=x$ and $aX=y$, we have

$$x : y :: \frac{d'}{2} : \left(\frac{d'}{2} - x\right)$$

$$\therefore y=\frac{2}{d'}\cdot\left\{\left(\frac{d'}{2} - x\right)x\right\} \qquad (68)$$

This is the equation of a parabola with vertex at n, half way between H K and the neutral axis. Hence the base of each prism is composed of parts of two symmetrical parabolas.

Areas of the bases. Since the area of a parabola is *two-thirds* of the circumscribing rectangle, the area of each base

$$=\tfrac{2}{3}\,(GO \times n\,v)$$

But $n\,v=\frac{1}{2}\,R'T'$ and $R'T'=\frac{1}{2}\,HK$

$$\therefore\ n\,v=\tfrac{1}{4}\,HK=\tfrac{1}{4}\,d'$$

$$\therefore\ \text{Area} = \tfrac{2}{3}\cdot\frac{d'}{2}\cdot\frac{d'}{4}=\frac{1}{12}d'^2$$

The centres of gravity of these bases (and consequently of the prisms) are at C and C′, and the distance

$$CC'=\tfrac{1}{2}\,d'$$

Hence the moment of resistance of the fibres about C or C′ is

$$M = \frac{1}{24} S\, d'^3 \qquad (69)$$

(S=height of prism or stress on external fibres at G and P.)

Corollary. To compare the resistance of the beam in this position with its resistance when lying flat :

Let d=side of the square as L G.

Then $d' = d\sqrt{2}$ and eq. (69) becomes

$$M = \frac{1}{12} S\, d^3 \sqrt{2} = \frac{1}{6\sqrt{2}} S\, d^3 \qquad (70)$$

Comparing this with eq. (66), we see that the beam offers greater resistance when flat in the proportion of

$$\frac{1}{6} : \frac{1}{6\sqrt{2}}$$

In solving these problems with diagonally placed beams, place the above value of M equal to the moment of the weight as before.

III. Let us apply this method to a T beam. Take for example the cast-iron ⊥ beam, calculated in part on p. 257 Rankine's Civil Engineering, in which the area of the flange = $\frac{2}{3}$ that of the web. Assume the flange to be 6 inches by 1 inch, and the web to be 5 inches by .8 of an inch, and draw a figure of the cross section to scale (Fig. 69).

1. As the top and bottom of this section is not symmetrical, it is necessary to find the position of the neutral axis, which is no longer at the half-depth. This may be done by calculation, or by a simple mechanical process as follows:

The centre of gravity of the cross section, since this last is symmetrical with regard to the vertical line through the middle of the web, must lie on this line. Cut accurately the figure of the cross section out of card board, or tin, or good paper, and suspend it freely by one end of the flange, suspending also from the same point a plummet. Mark the line of the plummet on the card board, and the centre of gravity being

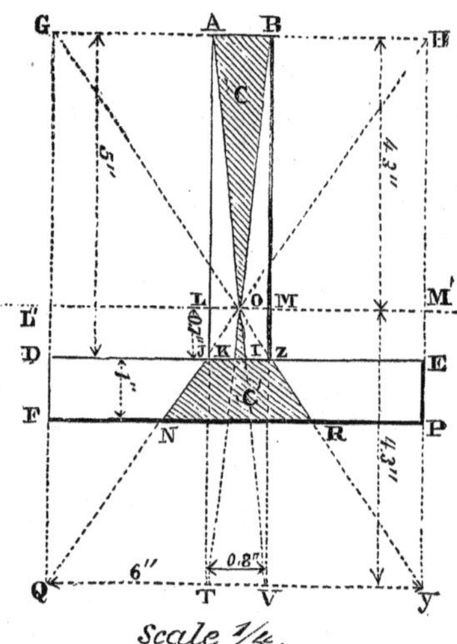

Scale 1/4.

FIG. 69.

on this line and also on the middle line of the web, will be at their intersection O. By measurement this point was found to be distant from A B four and three-tenths inches (4 . 3″), which is also the value by calculations. The line L M drawn through this point is the neutral axis.

2. To determine the bases of the prisms of stress. On the upper side the base is the triangle O A B, if the altitude be taken equal to the stress on the fibres along A B. For if L′ M′ H G were the upper half section, O G H would be the base of the prism and O A B is less than O G H, in the same proportion that L M A B, the real half section, is less than L′ M′ H G. Hence (if the upper be the compressed side) the total compressive force is equal to the prism erected on A O B, with the height equal to the unit stress at A B.

Below the axis L M.—For convenience we should have the height of the tension prism equal to that of the compression one, and the base must be determined under this condition. Complete the large rectangle G H Q Y, making the distance of Q Y below O=4.3 inches. Draw O Q and O Y and the shaded trapezoid J N R Z cut out on the flange by these lines, will evidently be the portion of the base due to the flange. Having prolonged the lines of the web to T and V, draw O T and O V, and then the shaded triangle O K I will be that part of the base due to the portion (L Z) of the web below the neutral axis.

The total tensile and compressive forces being always equal, and the height of the prisms having been assumed, each $=S'=$ stress on fibres at the distance of A B from O, the bases of these prisms must be equal also. This necessary equality between the area of O A B and that of O K I+J N R Z, affords a means of testing the accuracy of our work in finding the position of O.

3. Area of base O A B. This is,

$$=\frac{1}{2}(LA \times AB) = \frac{1}{2}(4.3 \times .8) = 1.72 \text{ sq. inches.}$$

4. To determine the distance C C′ (Fig. 69) between the centres of gravity of the prisms, which distance is the lever-arm to be used when one of these points is taken as the centre of moments. These centres of gravity (C and C′) can be readily determined by means similar to those employed in finding the centre

of gravity of the cross-section itself. Thus, cut the shaded areas (Fig. 69) out of card board or paper, and suspending each of them from two points in succession, draw vertical lines through the points of suspension. The intersection of these two lines gives the centre of gravity. In the present case they may be so simply obtained by calculation, that we adopt that method. The centre of gravity of A O B is $= \frac{2}{3}$ the distance from O to A B or O C$= \frac{2}{3}$ (4 . 3) $= 2 . 87$ inches. As to the shaded part below O, by using the ordinary formula for the centre of gravity and taking moments around O, we find the distance

$$OC' = \left\{ \frac{ONR \times \frac{2}{3}(1.7) - (OJK + OIZ)\frac{2}{3}(.7)}{ONR - 2\,(OJK)} \right\}$$

$$= \frac{2.2764 - .01388}{2.0145 - .2975} = 1.318 \text{ inch.}$$

Hence the distance

$$CC' = OC + OC' = 2.87 + 1.318 = 4.18 \text{ ins.}$$

Hence, since S′ is the height of the

prisms, the moment of resistance of the fibres is

$$M=4.18\times 1.72.\ S'=7.19\ S'$$

If it be desired to have M, not in terms of S′, the stress along A B, but of S″ the stress on the lowermost fibres (at F P), we have since the stresses increase directly with the distance from O,

$$S' : S'' :: L'G : L'F :: 4.3'' : 1.7''$$
$$\therefore\ S'=2.53.\ S''\ \text{and}\ M=18.19.\ S''$$

IV. As an illustration of the great saving of labor sometimes effected by this process, take the steel rail now widely used in England, the cross section of which is given to scale in (Fig. 70). The determination of the moment here by calculation would be long and tedious. The dimensions of the cross section are given on the figure.

1. The centre of gravity O of the cross section is found by making a template as in the last case, and suspending it freely by a corner. The vertical through the point of suspension intersects A X

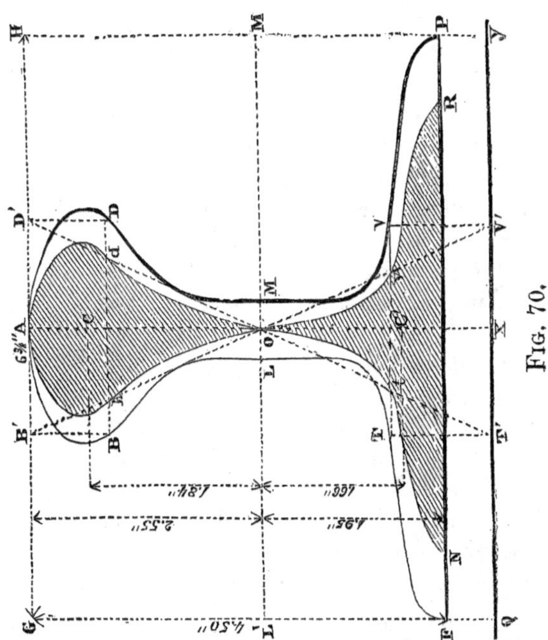

Fig. 70.

at O, which is 2.55 inches below A. Through this point draw L M, the neutral axis.

2. The bases of the prisms of stress

are determined by points as in example II.

Lay off O X = 2.55 inches. Draw the rectangle G H Q Y. Assume the height of the prisms to be the stress in the fibre at A. Then proceeding as in example II., the line of the base corresponding to any row of fibres, as B D, is *b d*. Obtain any number of points in the same way as *b* and *d*, and through these points draw a curve bounding the shaded figure A *b* O *d*. Similarly below the neutral axis, *t v* is the line in the base of the stress prism corresponding to T V, and the shaded figure, O N R, is that base where the height is taken equal to the unit stress at A. The equality of the bases in area is the test of accuracy.

3. To determine these areas. The simplest plan in the present case is first to find the area of the cross section itself. This is done as follows: The rail in question weighed 84 lbs. per yard, and the steel, of which it was made weighed, .277 lb. per cubic inch. Hence

if A=area of cross section in square inches

(36 . A) .277=84 .·. A=8.4 sq. inches.

Now cut out of the same card board, or paper, templates of the two shaded parts in the diagram, and also of the cross section itself, and weigh them. The ratio of the weights will equal that of the areas.

The comparison of weights may be readily made by means of a suspended wire, which may serve as a temporary balance, the templates to be compared being stuck on the opposite ends, and one or both moved until the wire is evenly balanced. The weights of the templates being inversely as their distances from the point of suspension of the wire, their areas will be in the same proportion. The areas of the prisms in the case before us were found to be equal each to 2.49 square inches.

4. The centres of gravity of these bases are found in the same way as those of the cross section itself. The point

C was thus found to be 1.84 inches above. O and C′ to be 1.66 inches below it. Hence the distance

$$C\,C'=3.5 \text{ inches.}$$

Therefore, finally, if S'=unit stress at A, the moment of resistance is,

$$M=2.49\times 3.5\times S'=8.715.\ S'$$

If we desire M in terms of S'' (stress at F P), we have

$$S' : S'' :: 1.84 : 1.66$$

$$\therefore\ S'=1.11\ S'' \quad \therefore\ M=9.67\ S''$$

V. A circular cross section (Fig. 71). Here the neutral axis of course=L M, passing through the centre. Draw the circumscribing rectangle G Y and obtain the points t, b, v, d, &c., in the curve bounding the base, as heretofore. Through the points so found draw the curves.

Determine the areas of the bases of the stress prisms by comparing them with the half-square G L M H. Thus, if a template is not to the surface of one of these beams it will just equal in

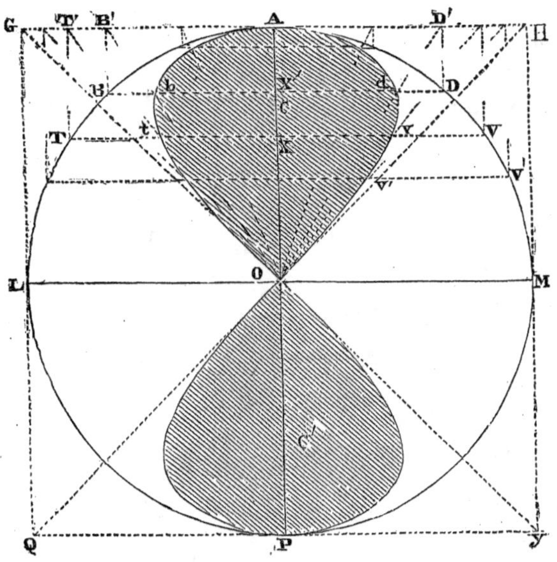

Fig. 71.

weight the template cut to the surface A G L O *t b* A, or, in other words, the shaded surface A *b* O *d* A is just one-third of the rectangle G M, or $\frac{1}{3}.\frac{d^2}{2}=\frac{1}{6}d^2$.

The centres of gravity C and C′ are

found as before by means of templates. In this way it was found that

$$O\,C = O\,C' = .587\,(O\,A) = .587 . \frac{d}{2}$$

$$\therefore\; C\,C' = .587\,d$$

The height of the prisms being S (= stress at A or F) the moment of resistance is,

$$M = S . \frac{d^2}{6} (.587\,d) = .0978\,d^3\,S$$

(The accurate value by calculation is $M = .0982\,d^3\,S$.)

Note.—The curve of the base of the prisms is a *lemniscate.* To find its equation we have in the triangles $O\,b\,X'$ and $O\,B'\,A$,

$$b\,X' : A\,B' (= B\,X') :: O\,X' : O\,A$$

And taking the vertical axis as that of X, and the horizontal one, as that of Y, the origin being at O, and calling the coordinates of the circle x' and y', and those of the *lemniscate* x and y we have:

$$y : y' :: x : R, \text{ or } y : \sqrt{R^2 - x^2} :: x : R$$

$$\therefore\; R^2\,y^2 = x^2\,(R^2 - x^2).$$

**** Any book in this Catalogue sent free by mail on receipt of price.*

VALUABLE

SCIENTIFIC BOOKS,

PUBLISHED BY

D. VAN NOSTRAND,

23 MURRAY STREET AND 27 WARREN STREET,

NEW YORK.

FRANCIS. Lowell Hydraulic Experiments, being a selection from Experiments on Hydraulic Motors, on the Flow of Water over Weirs, in Open Canals of Uniform Rectangular Section, and through submerged Orifices and diverging Tubes. Made at Lowell, Massachusetts. By James B. Francis, C. E. 2d edition, revised and enlarged, with many new experiments, and illustrated with twenty-three copperplate engravings. 1 vol. 4to, cloth......................$15 00

ROEBLING (J. A.) Long and Short Span Railway Bridges. By John A. Roebling, C. E. Illustrated with large copperplate engravings of plans and views. Imperial folio, cloth............................ 25 00

CLARKE (T. C.) Description of the Iron Railway Bridge over the Mississippi River, at Quincy, Illinois. Thomas Curtis Clarke, Chief Engineer. Illustrated with 21 lithographed plans. 1 vol. 4to, cloth 7 50

TUNNER (P.) A Treatise on Roll-Turning for the Manufacture of Iron. By Peter Tunner. Translated and adapted by John B. Pearse, of the Pennylvania Steel Works, with numerous engravings wood cuts and folio atlas of plates. $10 00

ISHERWOOD (B. F.) Engineering Precedents for Steam Machinery. Arranged in the most practical and useful manner for Engineers. By B. F. Isherwood, Civil Engineer, U. S. Navy. With Illustrations. Two volumes in one. 8vo, cloth $2 50

BAUERMAN. Treatise on the Metallurgy of Iron, containing outlines of the History of Iron Manufacture, methods of Assay, and analysis of Iron Ores, processes of manufacture of Iron and Steel, etc., etc. By H. Bauerman. First American edition. Revised and enlarged, with an Appendix on the Martin Process for making Steel, from the report of Abram S. Hewitt. Illustrated with numerous wood engravings. 12mo, cloth 2 00

CAMPIN on the Construction of Iron Roofs. By Francis Campin. 8vo, with plates, cloth 2 00

COLLINS. The Private Book of Useful Alloys and Memoranda for Goldsmiths, Jewellers, &c. By James E. Collins. 18mo, cloth 75

CIPHER AND SECRET LETTER AND TELEGRAPHIC CODE, with Hogg's Improvements. The most perfect secret code ever invented or discovered. Impossible to read without the key. By C. S. Larrabee. 18mo, cloth. 1 00

COLBURN. The Gas Works of London. By Zerah Colburn, C. E. 1 vol 12mo, boards 60

CRAIG (B. F.) Weights and Measures. An account of the Decimal System, with Tables of Conversion for Commercial and Scientific Uses. By B. F. Craig, M.D. 1 vol. square 32mo, limp cloth 50

NUGENT. Treatise on Optics; or, Light and Sight, theoretically and practically treated; with the application to Fine Art and Industrial Pursuits. By E. Nugent. With one hundred and three illustrations. 12mo, cloth 2 00

FREE HAND DRAWING. A Guide to Ornamental Figure and Landscape Drawing. By an Art Student. 18mo, boards, ... 50

HOWARD. Earthwork Mensuration on the Basis of the Prismoidal Formulae. Containing simple and labor-saving method of obtaining Prismoidal contents directly from End Areas. Illustrated by Examples, and accompanied by Plain Rules for Practical Uses. By Conway R. Howard, C. E., Richmond, Va. Illustrated, 8vo, cloth.............. .... 1 50

GRUNER. The Manufacture of Steel. By M. L. Gruner. Translated from the French, by Lenox Smith, with an appendix on the Bessamer process in the United States, by the translator. Illustrated by Lithographed drawings and wood cuts. 8vo, cloth.. 3 50

AUCHINCLOSS. Link and Valve Motions Simplified. Illustrated with 37 wood-cuts, and 21 lithographic plates, together with a Travel Scale, and numerous useful Tables. By W. S. Auchincloss. 8vo, cloth.. 3 00

VAN BUREN. Investigations of Formulas, for the strength of the Iron parts of Steam Machinery. By J. D. Van Buren, Jr., C. E. Illustrated, 8vo, cloth. 2 00

JOYNSON. Designing and Construction of Machine Gearing. Illustrated, 8vo, cloth.................. 2 00

GILLMORE. Coignet Beton and other Artificial Stone. By Q. A. Gillmore, Major U. S. Corps Engineers. 9 plates, views, &c. 8vo, cloth.................... 2 50

SAELTZER. Treattse on Acoustics in connection with Ventilation. By Alexander Saeltzer, Architect. 12mo, cloth.................................. 2 00

THE EARTH'S CRUST. A handy Outline of Geology. By David Page. Illustrated, 18mo, cloth.... 75

DICTIONARY of Manufactures, Mining, Machinery, and the Industrial Arts. By George Dodd. 12mo, cloth... 2 00

BOW. A Treatise on Bracing, with its application to Bridges and other Structures of Wood or Iron. By Robert Henry Bow, C. E. 156 illustrations, 8vo, cloth... 1 50

GILLMORE (Gen. Q. A.) Treatise on Limes, Hydraulic Cements, and Mortars. Papers on Practical Engineering, U. S. Engineer Department, No. 9, containing Reports of numerous Experiments conducted in New York City, during the years 1858 to 1861, inclusive. By Q. A. Gillmore, Bvt. Maj -Gen., U. S. A., Major, Corps of Engineers. With numerous illustrations. 1 vol, 8vo, cloth............... $4 00

HARRISON. The Mechanic's Tool Book, with Practical Rules and Suggestions for Use of Machinists, Iron Workers, and others. By W. B. Harrison, associate editor of the "American Artisan." Illustrated with 44 engravings. 12mo, cloth............ 1 50

HENRICI (Olaus). Skeleton Structures, especially in their application to the Building of Steel and Iron Bridges. By Olaus Henrici. With folding plates and diagrams. 1 vol. 8vo, cloth................... 1 50

HEWSON (Wm.) Principles and Practice of Embanking Lands from River Floods, as applied to the Levees of the Mississippi. By William Hewson, Civil Engineer. 1 vol. 8vo, cloth....................... 2 00

HOLLEY (A. L.) Railway Practice. American and European Railway Practice, in the economical Generation of Steam, including the Materials and Construction of Coal-burning Boilers, Combustion, the Variable Blast, Vaporization, Circulation, Superheating, Supplying and Heating Feed-water, etc., and the Adaptation of Wood and Coke-burning Engines to Coal-burning; and in Permanent Way, including Road-bed, Sleepers, Rails, Joint-fastenings, Street Railways, etc., etc. By Alexander L. Holley, B. P. With 77 lithographed plates. 1 vol. folio, cloth.... 12 00

KING (W. H.) Lessons and Practical Notes on Steam, the Steam Engine, Propellers, etc., etc., for Young Marine Engineers, Students, and others. By the late W. H. King, U. S. Navy. Revised by Chief Engineer J. W. King, U. S. Navy. Twelfth edition, enlarged. 8vo, cloth. 2 00

MINIFIE (Wm.) Mechanical Drawing. A Text-Book of Geometrical Drawing for the use of Mechanics and Schools, in which the Definitions and Rules of Geometry are familiarly explained; the Practical Problems are arranged, from the most simple to the more complex, and in their description technicalities are avoided as much as possible. With illustrations for Drawing Plans, Sections, and Elevations of Railways and Machinery; an Introduction to Isometrical Drawing, and an Essay on Linear Perspective and Shadows. Illustrated with over 200 diagrams engraved on steel. By Wm. Minifie, Architect. Seventh edition. With an Appendix on the Theory and Application of Colors. 1 vol. 8vo, cloth........... $4 00

"It is the best work on Drawing that we have ever seen, and is especially a text-book of Geometrical Drawing for the use of Mechanics and Schools. No young Mechanic, such as a Machinists, Engineer, Cabinet-maker, Millwright, or Carpenter, should be without it."—*Scientific American.*

—— Geometrical Drawing. Abridged from the octavo edition, for the use of Schools. Illustrated with 48 steel plates. Fifth edition. 1 vol. 12mo, cloth.... 2 00

STILLMAN (Paul.) Steam Engine Indicator, and the Improved Manometer Steam and Vacuum Gauges—their Utility and Application. By Paul Stillman. New edition. 1 vol. 12mo, flexible cloth........... 1 00

SWEET (S. H.) Special Report on Coal; showing its Distribution, Classification, and cost delivered over different routes to various points in the State of New York, and the principal cities on the Atlantic Coast. By S. H. Sweet. With maps, 1 vol. 8vo, cloth..... 3 00

WALKER (W. H.) Screw Propulsion. Notes on Screw Propulsion: its Rise and History. By Capt. W. H. Walker, U. S. Navy. 1 vol. 8vo, cloth..... 75

WARD (J. H.) Steam for the Million. A popular Treatise on Steam and its Application to the Useful Arts, especially to Navigation. By J. H. Ward, Commander U. S. Navy. New and revised edition. 1 vol. 8vo, cloth................................ 1 00

WEISBACH (Julius). Principles of the Mechanics of Machinery and Engineering. By Dr. Julius Weisbach, of Freiburg. Translated from the last German edition. Vol. I., 8vo, cloth.. .... 10 00

DIEDRICH. The Theory of Strains, a Compendium for the calculation and construction of Bridges, Roofs, and Cranes, with the application of Trigonometrical Notes, containing the most comprehensive information in regard to the Resulting strains for a permanent Load, as also for a combined (Permanent and Rolling) Load. In two sections, adadted to the requirements of the present time. By John Diedrich, C. E. Illustrated by numerous plates and diagrams. 8vo, cloth.................................... 5 00

WILLIAMSON (R. S.) On the use of the Barometer on Surveys and Reconnoissances. Part I. Meteorology in its Connection with Hypsometry. Part II. Barometric Hypsometry. By R. S. Wiliamson, Bvt Lieut.-Col. U. S. A., Major Corps of Engineers. With Illustrative Tables and Engravings. Paper No. 15, Professional Papers, Corps of Engineers. 1 vol. 4to, cloth.................................... 15 00

POOK (S. M.) Method of Comparing the Lines and Draughting Vessels Propelled by Sail or Steam. Including a chapter on Laying off on the Mould-Loft Floor. By Samuel M. Pook, Naval Constructor. 1 vol. 8vo, with illustrations, cloth........... 5 00

ALEXANDER (J. H.) Universal Dictionary of Weights and Measures, Ancient and Modern, reduced to the standards of the United States of America. By J. H. Alexander. New edition, enlarged. 1 vol. 8vo, cloth.................................... 3 50

WANKLYN. A Practical Treatise on the Examination of Milk, and its Derivatives, Cream, Butter and Cheese. By J. Alfred Wanklyn, M. R. C. S., 12mo, cloth.................................... 1 00

RICHARDS' INDICATOR. A Treatise on the Richards Steam Engine Indicator, with an Appendix by F. W. Bacon, M. E. 18mo, flexible, cloth.......... 1 00

POPE. Modern Practice of the Electric Telegraph. A Hand Book for Electricians and operators. By Frank L. Pope. Eighth edition, revised and enlarged, and fully illustrated. 8vo, cloth........................ $2.00

"There is no other work of this kind in the English language that contains in so small a compass so much practical information in the application of galvanic electricity to telegraphy. It should be in the hands of every one interested in telegraphy, or the use of Batteries for other purposes."

MORSE. Examination of the Telegraphic Apparatus and the Processes in Telegraphy. By Samuel F. Morse, LL.D., U. S. Commissioner Paris Universal Exposition, 1867. Illustrated, 8vo, cloth.......... $2 00

SABINE. History and Progress of the Electric Telegraph, with descriptions of some of the apparatus. By Robert Sabine, C. E. Second edition, with additions, 12mo, cloth............................... 1 25

BLAKE. Ceramic Art. A Report on Pottery, Porcelain, Tiles, Terra Cotta and Brick. By W. P. Blake, U. S. Commissioner, Vienna Exhibition, 1873. 8vo, cloth... 2 00

BENET. Electro-Ballistic Machines, and the Schultz Chronoscope. By Lieut.-Col. S. V. Benet, Captain of Ordnance, U. S. Army. Illustrated, second edition, 4to, cloth.................................. 3 00

MICHAELIS. The Le Boulenge Chronograph, with three Lithograph folding plates of illustrations. By Brevet Captain O. E. Michaelis, First Lieutenant Ordnance Corps, U. S. Army, 4to, cloth.......... 3 00

ENGINEERING FACTS AND FIGURES An Annual Register of Progress in Mechanical Engineering and Construction. for the years 1863, 64, 65, 66, 67, 68. Fully illustrated, 6 vols. 18mo, cloth, $2.50 per vol., each volume sold separately.............

HAMILTON. Useful Information for Railway Men. Compiled by W. G. Hamilton, Engineer. Fifth edition, revised and enlarged, 562 pages Pocket form. Morocco, gilt.................................... 2 00

STUART. The Civil and Military Engineers of America. By Gen. C. B. Stuart. With 9 finely executed portraits of eminent engineers, and illustrated by engravings of some of the most important works constructed in America. 8vo, cloth.................. $5 00

STONEY. The Theory of Strains in Girders and similar structures, with observations on the application of Theory to Practice, and Tables of Strength and other properties of Materials. By Bindon B. Stoney, B. A. New and revised edition, enlarged, with numerous engravings on wood, by Oldham. Royal 8vo, 664 pages. Complete in one volume. 8vo, cloth...... 12 50

SHREVE. A Treatise on the Strength of Bridges and Roofs. Comprising the determination of Algebraic formulas for strains in Horizontal, Inclined or Rafter, Triangular, Bowstring, Lenticular and other Trusses, from fixed and moving loads, with practical applications and examples, for the use of Students and Engineers. By Samuel H. Shreve, A. M., Civil Engineer. 87 wood cut illustrations. 8vo, cloth.............. 5 00

MERRILL. Iron Truss Bridges for Railroads. The method of calculating strains in Trusses, with a careful comparison of the most prominent Trusses, in reference to economy in combination, etc., etc. By Brevet. Col. William E. Merrill, U S. A., Major Corps of Engineers, with nine lithographed plates of Illustrations. 4to, cloth.......................... 5 00

WHIPPLE. An Elementary and Practical Treatise on Bridge Building. An enlarged and improved edition of the author's original work. By S. Whipple, C. E., inventor of the Whipple Bridges, &c. Illustrated 8vo, cloth.................................... 4 00

THE KANSAS CITY BRIDGE. With an account of the Regimen of the Missouri River, and a description of the methods used for Founding in that River. By O. Chanute, Chief Engineer, and George Morrison, Assistant Engineer. Illustrated with five lithographic views and twelve plates of plans. 4to, cloth, 6 00

MAC CORD. A Practical Treatise on the Slide Valve by Eccentrics, examining by methods the action of the Eccentric upon the Slide Valve, and explaining the Practical processes of laying out the movements, adapting the valve for its various duties in the steam engine. For the use of Engineers, Draughtsmen, Machinists, and Students of Valve Motions in general. By C. W. Mac Cord, A. M., Professor of Mechanical Drawing, Stevens' Institute of Technology, Hoboken, N. J. Illustrated by 8 full page copper-plates. 4to, cloth.................................. $4 00

KIRKWOOD. Report on the Filtration of River Waters, for the supply of cities, as practised in Europe, made to the Board of Water Commissioners of the City of St. Louis. By James P. Kirkwood. Illustrated by 30 double plate engravings. 4to, cloth, 15 00

PLATTNER. Manual of Qualitative and Quantitative Analysis with the Blow Pipe. From the last German edition, revised and enlarged. By Prof. Th. Richter, of the Royal Saxon Mining Academy. Translated by Prof. H. B. Cornwall, Assistant in the Columbia School of Mines, New York assisted by John H. Caswell. Illustrated with 87 wood cuts, and one lithographic plate. Second edition, revised, 560 pages, 8vo, cloth.................................. 7 50

PLYMPTON. The Blow Pipe. A Guide to its Use in the Determination of Salts and Minerals. Compiled from various sources, by George W. Plympton, C. E. A. M., Professor of Physical Science in the Polytechnic Institute, Brooklyn, New York, 12mo, cloth.. 1 50

PYNCHON. Introduction to Chemical Physics, designed for the use of Academies, Colleges and High Schools. Illustrated with numerous engravings, and containing copious experiments with directions for preparing them. By Thomas Ruggles Pynchon, M. A., Professor of Chemistry and the Natural Sciences, Trinity College, Hartford New edition, revised and enlarged, and illustrated by 269 illustrations on wood. Crown, 8vo. cloth...................... 3 00

ELIOT AND STORER. A compendious Manual of Qualitative Chemical Analysis. By Charles W. Eliot and Frank H. Storer. Revised with the Co-operation of the authors. By William R. Nichols, Professor of Chemistry in the Massachusetts Institute of Technology. Illustrated, 12mo, cloth....... $1 50

RAMMELSBERG. Guide to a course of Quantitative Chemical Analysis, especially of Minerals and Furnace Products. Illustrated by Examples. By C. F. Rammelsberg. Translated by J. Towler, M. D. 8vo, cloth.......... 2 25

EGLESTON. Lectures on Descriptive Mineralogy, delivered at the School of Mines, Columbia College. By Professor T. Egleston. Illustrated by 34 Lithographic Plates. 8vo, cloth.......... 4 50

JACOB. On the Designing and Construction of Storage Reservoirs, with Tables and Wood Cuts representing Sections, &c., 18mo, boards.......... 50

WATT'S Dictionary of Chemistry. New and Revised edition complete in 6 vols. 8vo cloth, $62.00. Supplementary volume sold separately. Price, cloth... 9 00

RANDALL. Quartz Operators Hand-Book. By P. M. Randall. New edition, revised and enlarged, fully illustrated. 12mo, cloth.......... 2 00

SILVERSMITH. A Practical Hand-Book for Miners, Metallurgists, and Assayers, comprising the most recent improvements in the disintegration amalgamation, smelting, and parting of the Precious ores, with a comprehensive Digest of the Mining Laws. Greatly augmented, revised and corrected. By Julius Silversmith. Fourth edition. Profusely illustrated. 12mo, cloth.......... 3 00

THE USEFUL METALS AND THEIR ALLOYS, including Mining Ventilation, Mining Jurisprudence, and Metallurgic Chemistry employed in the conversion of Iron, Copper, Tin, Zinc, Antimony and Lead ores, with their applications to the Industrial Arts. By Scoffren, Truan, Clay, Oxland, Fairbairn, and others. Fifth edition, half calf.......... 3 75

JOYNSON. The Metals used in construction, Iron, Steel, Bessemer Metal, etc., etc. By F. H. Joynson. Illustrated, 12mo, cloth.......................... $0 75

VON COTTA. Treatise on Ore Deposits. By Bernhard Von Cotta, Professor of Geology in the Royal School of Mines, Freidberg, Saxony. Translated from the second German edition, by Frederick Prime, Jr., Mining Engineer, and revised by the author, with numerous illustrations. 8vo, cloth....... 4 00

GREENE. Graphical Method for the Analysis of Bridge Trusses, extended to continuous Girders and Draw Spans. By C. E. Greene, A. M., Prof. of Civil Engineering, University of Michigan. Illustrated by 3 folding plates, 8vo, cloth.......................... 2 00

BELL. Chemical Phenomena of Iron Smelting. An experimental and practical examination of the circumstances which determine the capacity of the Blast Furnace, The Temperature of the air, and the proper condition of the Materials to be operated upon. By I. Lowthian Bell. 8vo, cloth........... 6 00

ROGERS. The Geology of Pennsylvania. A Government survey, with a general view of the Geology of the United States, Essays on the Coal Formation and its Fossils, and a description of the Coal Fields of North America and Great Britain. By Henry Darwin Rogers, late State Geologist of Pennsylvania, Splendidly illustrated with Plates and Engravings in the text. 3 vols., 4to, cloth, with Portfolio of Maps. 30 00

BURGH. Modern Marine Engineering, applied to Paddle and Screw Propulsion. Consisting of 36 colored plates, 259 Practical Wood Cut Illustrations, and 403 pages of descriptive matter, the whole being an exposition of the present practice of James Watt & Co., J. & G. Rennie, R. Napier & Sons, and other celebrated firms, by N. P. Burgh, Engineer, thick 4to, vol., cloth, $25.00; half mor........ 30 00

CHURCH. Notes of a Metallurgical Journey in Europe. By J. A. Church, Engineer of Mines, 8vo, cloth..... 2 00

BOURNE. Treatise on the Steam Engine in its various applications to Mines, Mills, Steam Navigation, Railways, and Agriculture, with the theoretical investigations respecting the Motive Power of Heat, and the proper proportions of steam engines. Elaborate tables of the right dimensions of every part, and Practical Instructions for the manufacture and management of every species of Engine in actual use. By John Bourne, being the ninth edition of "A Treatise on the Steam Engine," by the "Artizan Club." Illustrated by 38 plates and 546 wood cuts. 4to, cloth.......................................$15 00

STUART. The Naval Dry Docks of the United States. By Charles B. Stuart late Engineer-in-Chief of the U. S. Navy. Illustrated with 24 engravings on steel. Fourth edition, cloth.................... 6 00

ATKINSON. Practical Treatises on the Gases met with in Coal Mines. 18mo, boards................ 50

FOSTER. Submarine Blasting in Boston Harbor, Massachusetts. Removal of Tower and Corwin Rocks. By J. G. Foster, Lieut.-Col. of Engineers, U. S. Army. Illustrated with seven plates, 4to, cloth.. 3 50

BARNES Submarine Warfare, offensive and defensive, including a discussion of the offensive Torpedo System, its effects upon Iron Clad Ship Systems and influence upon future naval wars. By Lieut.-Commander J. S. Barnes, U. S. N., with twenty lithographic plates and many wood cuts. 8vo, cloth..... 5 00

HOLLEY. A Treatise on Ordnance and Armor, embracing descriptions, discussions, and professional opinions concerning the materials, fabrication, requirements, capabilities, and endurance of European and American Guns, for Naval, Sea Coast, and Iron Clad Warfare, and their Rifling, Projectiles, and Breech-Loading; also, results of experiments against armor, from official records, with an appendix referring to Gun Cotton, Hooped Guns, etc., etc. By Alexander L. Holley, B. P., 948 pages, 493 engravings, and 147 Tables of Results, etc., 8vo, half roan. 10 00

SIMMS. A Treatise on the Principles and Practice of Levelling, showing its application to purposes of Railway Engineering and the Construction of Roads, &c. By Frederick W. Simms, C. E. From the 5th London edition, revised and corrected, with the addition of Mr. Laws's Practical Examples for setting out Railway Curves. Illustrated with three Lithographic plates and numerous wood cuts. 8vo, cloth. $2 50

BURT. Key to the Solar Compass, and Surveyor's Companion; comprising all the rules necessary for use in the field; also description of the Linear Surveys and Public Land System of the United States, Notes on the Barometer, suggestions for an outfit for a survey of four months, etc. By W. A. Burt, U. S. Deputy Surveyor. Second edition. Pocket book form, tuck.................................... 2 50

THE PLANE TABLE. Its uses in Topographical Surveying, from the Papers of the U. S. Coast Survey. Illustrated, 8vo, cloth...................... 2 00

"This work gives a description of the Plane Table, employed at the U. S. Coast Survey office, and the manner of using it."

JEFFER'S. Nautical Surveying. By W. N. Jeffers, Captain U. S. Navy. Illustrated with 9 copperplates and 31 wood cut illustrations. 8vo, cloth.......... 5 00

CHAUVENET. New method of correcting Lunar Distances, and improved method of Finding the error and rate of a chronometer, by equal altitudes. By W. Chauvenet, LL.D. 8vo, cloth................ 2 00

BRUNNOW. Spherical Astronomy. By F. Brunnow, Ph. Dr. Translated by the author from the second German edition. 8vo, cloth...................... 6 50

PEIRCE. System of Analytic Mechanics. By Benjamin Peirce. 4to, cloth.......................... 10 00

COFFIN. Navigation and Nautical Astronomy. Prepared for the use of the U. S. Naval Academy. By Prof. J. H. C. Coffin. Fifth edition. 52 wood cut illustrations. 12mo, cloth.......................... 3 50

CLARK. Theoretical Navigation and Nautical Astronomy. By Lieut. Lewis Clark, U. S. N. Illustrated with 41 wood cuts. 8vo, cloth.................... $3 00

HASKINS. The Galvanometer and its Uses. A Manual for Electricians and Students. By C. H. Haskins. 12mo, pocket form, morocco. (In press).....

GOUGE. New System of Ventilation, which has been thoroughly tested, under the patronage of many distinguished persons. By Henry A. Gouge. With many illustrations. 8vo, cloth...................... 2 00

BECKWITH. Observations on the Materials and Manufacture of Terra-Cotta, Stone Ware, Fire Brick, Porcelain and Encaustic Tiles, with remarks on the products exhibited at the London International Exhibition, 1871. By Arthur Beckwith, C. E. 8vo, paper.. 60

MORFIT. A Practical Treatise on Pure Fertilizers, and the chemical conversion of Rock Guano, Marlstones, Coprolites, and the Crude Phosphates of Lime and Alumina generally, into various valuable products. By Campbell Morfit, M.D., with 28 illustrative plates, 8vo, cloth.................................... 20 00

BARNARD. The Metric System of Weights and Measures. An address delivered before the convocation of the University of the State of New York, at Albany, August, 1871. By F. A. P. Barnard, LL.D., President of Columbia College, New York. Second edition from the revised edition, printed for the Trustees of Columbia College. Tinted paper, 8vo, cloth 3 00

—— Report on Machinery and Processes on the Industrial Arts and Apparatus of the Exact Sciences. By F. A. P. Barnard, LL.D. Paris Universal Exposition, 1867. Illustrated, 8vo, cloth.............. 5 00

ALLAN. Theory of Arches. By Prof. W. Allan, formerly of Washington & Lee University, 18mo, b'rds 50

MYER. Manual of Signals, for the use of Signal officers in the Field, and for Military and Naval Students, Military Schools, etc. A new edition enlarged and illustrated. By Brig. General Albert J. Myer, Chief Signal Officer of the army, Colonel of the Signal Corps during the War of the Rebellion. 12mo, 48 plates, full Roan.................................. $5 00

WILLIAMSON. Practical Tables in Meteorology and Hypsometry, in connection with the use of the Barometer. By Col. R. S. Williamson, U. S. A. 4to, cloth.. 2 50

CLEVENGER. A Treatise on the Method of Government Surveying, as prescribed by the U. S. Congress and Commissioner of the General Land Office, with complete Mathematical, Astronomical and Practical Instructions for the Use of the United States Surveyors in the Field. By S. R. Clevenger, Pocket Book Form, Morocco............................ 2 50

PICKERT AND METCALF. The Art of Graining. How Acquired and How Produced, with description of colors, and their application. By Charles Pickert and Abraham Metcalf. Beautifully illustrated with 42 tinted plates of the various woods used in interior finishing. Tinted paper, 4to, cloth............... 10 00

HUNT. Designs for the Gateways of the Southern Entrances to the Central Park. By Richard M. Hunt. With a description of the designs. 4to. cloth...... 5 00

LAZELLE. One Law in Nature. By Capt. H. M. Lazelle, U. S. A. A new Corpuscular Theory, comprehending Unity of Force, Identity of Matter, and its Multiple Atom Constitution, applied to the Physical Affections or Modes of Energy. 12mo, cloth... 1 50

CORFIELD. Water and Water Supply. By W. H. Corfield, M. A. M. D., Professor of Hygiene and Public Health at University College, London. 18mo, boards.. 50

BOYNTON. History of West Point, its Military Importance during the American Revolution, and the Origin and History of the U. S. Military Academy. By Bvt. Major C. E. Boynton, A.M., Adjutant of the Military Academy. Second edition, 416 pp. 8vo, printed on tinted paper, beautifully illustrated with 36 maps and fine engravings, chiefly from photographs taken on the spot by the author. Extra cloth.. $3 50

WOOD. West Point Scrap Book, being a collection of Legends, Stories, Songs, etc., of the U. S. Military Academy. By Lieut. O. E. Wood, U. S. A. Illustrated by 69 engravings and a copperplate map. Beautifully printed on tinted paper. 8vo, cloth..... 5 00

WEST POINT LIFE. A Poem read before the Dialectic Society of the United States Military Academy. Illustrated with Pen-and-Ink Sketches. By a Cadet. To which is added the song, "Benny Havens, oh!" oblong 8vo, 21 full page illustrations, cloth......... 2 50

GUIDE TO WEST POINT and the U. S. Military Academy, with maps and engravings, 18mo, blue cloth, flexible.................................... 1 00

HENRY. Military Record of Civilian Appointments in the United States Army. By Guy V. Henry, Brevet Colonel and Captain First United States Artillery, Late Colonel and Brevet Brigadier General, United States Volunteers. Vol. 1 now ready. Vol. 2 in press. 8vo, per volume, cloth..................... 5 00

HAMERSLY. Records of Living Officers of the U. S. Navy and Marine Corps. Compiled from official sources. By Lewis B. Hamersly, late Lieutenant U. S. Marine Corps. Revised edition, 8vo, cloth... 5 00

MOORE. Portrait Gallery of the War. Civil, Military and Naval. A Biographical record, edited by Frank Moore. 60 fine portraits on steel. Royal 8vo, cloth.. 6 00

www.ingramcontent.com/pod-product-compliance
Lightning Source LLC
LaVergne TN
LVHW021418110826
845150LV00007B/1977

* 9 7 8 1 4 2 5 5 0 9 6 0 6 *